AMONG THE MUSHROOMS
A Guide for Beginners

AMONG THE MUSHROOMS
A Guide for Beginners

Ellen M. Dallas

and

Caroline A. Burgin

MJP Publishers

Chennai   New Delhi   Tirunelveli

ISBN 978-81-8094-203-7

MJP Publishers

New No. 5 Muthu Kalathy Street,
Triplicane,
Chennai 600 005

MJP 183

© Publishers, 2013

Publisher : J.C. Pillai

This book has been published in good faith that the work of the author is original. All efforts have been taken to make the material error-free. However, the author and publisher disclaim responsibility for any inadvertent errors.

"*Have you not seen in the woods on a late autumn morning a poor fungus or mushroom—a plant without any solidity, nay, that seemed nothing but a soft mush or jelly—by its constant total and inconceivably gentle pushing, manage to break its way up through the frosty ground, and actually to lift a hard crust on its head? It is the symbol of the power of kindness.*"

Emerson

PREFACE

The books which have been consulted in the preparation of this work are, "British Fungi," by Rev. John Stevenson; "British Fungus-Flora," by George Massee; "Mushrooms and their Uses," and "Boleti of the United States," by Professor Charles H. Peck, State Botanist of New York; "Moulds, Mildew and Mushrooms," by Professor L. M. Underwood; and a pamphlet by Mr. C. G. Lloyd, entitled "The Volvæ of the United States."

No attempt has been made to do more than to put in popular language the statements of experienced botanists, and so to arrange the matter as to aid beginners in their work.

Thanks are due to Mr. Harold Wingate for his suggestions and corrections of the manuscript; to Mr. C. G. Lloyd for permission to print from his photographs; to Miss Laura C. Detwiller for her paintings from nature, which have been here reproduced; and also to Mrs. Harrison Streeter and Miss Mary W. Nichols for their encouragement of the undertaking and suggestions in furtherance of its success.

CONTENTS

Chapter 1

INTRODUCTION

This book is intended for those who, though ignorant on the subject, desire to know something about mushrooms. The first question which such an one asks upon finding a mushroom is, "What is its name?" If there is no one near to tell him, then follows the second inquiry, "How can I find it out for myself?" If wild flowers were concerned, Gray's little book, "How the Plants Grow," could be used; and there is also Mrs. Dana's book on "The Wild Flowers," that has given so much pleasure. In the case of mushrooms, however, but one answer can be returned to all questions: "There is no American text-book on mushrooms, there is no manual for beginners."

There are many books on British fungi for students, but we want some popular work easy to understand, with no technical expressions.

This necessity for a simple guide-book has been felt by many. Let us give our own experience. We procured a list of works on fungi, and looked for some volume not too deep for our comprehension nor too costly for our purse. Among those we found were "Handbook for Students" (Taylor); "Edible and Poisonous Fungi" (Cooke), and a pamphlet by Professor Peck, "Mushrooms and Their Uses." This seemed to be the one that we could comprehend most easily, and so, armed with it, and another pamphlet by Professor Underwood, called "Suggestions to Collectors of Fleshy Fungi," which contained a simple key, we started out to make discoveries. We afterward procured some publications of Mr. C. G. Lloyd, which were of great assistance, and lastly a glossary published by the Boston Mycological Society, a necessary addition to our library.

We found Professor Peck's book was confined to edible mushrooms, and it soon became too limited to satisfy our craving for further knowledge—it incited a longing to know something of inedible fungi.

The rest is soon told. We were advised to get either a copy of Stevenson's "British Fungi" or of Massee's works. We did so, but found them too advanced to be readily used by the unlearned. Then the idea arose, How can we help others in their difficulties? This little book is the answer. It will not be of use to advanced students, they will only criticise and discover how much has been left unsaid; but the beginner is more easily satisfied with the extent of information gained, and if a taste for knowledge is encouraged the object of this book is attained.

This explanation will also account for the use of simple terms. We find a tiny fungus which looks like a brownish bird's nest, with some miniature eggs in it, or a shining white mushroom, and we are told its name in Latin; it is described in terms meaningless to the ignorant, we lose interest, and our attention flags. We began for pleasure and recreation, but it became irksome and fatiguing, and the subject which might have amused us and helped to pass many an idle hour is put aside and abandoned. Yet this study is a most fascinating one. We all long for pleasant subjects of thought in our leisure hours, and there can be nothing more diverting and absorbing than the investigation of the beautiful and familiar plants around us.

When we leave the bustling, noisy streets of a city and go into the quiet fields and woods the contrast is very great.

A walk for exercise alone is often dull and tiresome. We cannot be assured of pleasant companions, nor is there always a fine view or picturesque scenery to reward us during our strolls, but there are plants to be found and gathered, and when these fail us, then the bright-hued mushrooms may arrest our attention. The discovery of new specimens, the learning their names, the knowledge of their curious organizations, will all add an interest to our lives. It will inspire us with a love of nature, and open our eyes to many objects of which we have before been unobservant. Besides this it obliges us to be accurate. Our descriptions must be exact or they are of no use.

Let us imagine ourselves taking a stroll in the woods or down some shady lane, and see what we can find there.

The golden-rod and asters adorn the roadsides, the odors of the sweet gale and scented fern are wafted gratefully to our senses as we pass along the lanes, and there, among the fallen leaves, at the very edge of the woods, peers out a bright yellow mushroom, brighter from the contrast to the dead leaves around, and then another, close by, and then a shining white cap; further on a mouse-colored one, gray, and silky in texture. What a contrast of colors. What are they? By what names shall we call them?

Let us first carefully dig up the yellow one. We have brought a basket and trowel, and can examine them thoroughly. We must dig down deep so as not to break off the stem. There is a ring or collar around it near the top. There is a bulb at the base, with some slight membrane attached. The cap is orange color, almost smooth, covered with a few spots like warts, and there are some lines on the margin. The gills are not attached to the stem, and are white with a creamy hue. The stem is also white, tinged faintly with yellow. We will take a penknife and divide it into halves, cutting straight through the stem and cap. We find the stem is filled with a spongy substance, and we can now see more clearly the position of the gills. Our specimen measures 2 inches across the cap, and the stem is 2 or 3 inches long. It is an Amanita, resembling the Fly Amanita, which we will probably soon discover. Our fungus is Frost's Amanita, named after the botanist who first placed it on the list, Frost. It is not among the British fungi. It is American.

Now let us dig up the shining white one. It is much larger than the yellow fungus, handsome, pure-looking, with a rather slender stem. The cap is nearly 4 inches across, the flesh is white. The stem is long, solid, with a bulbous base. There is a wide, loose ring high up on the stem. The membrane around the base is large and thick. The stem is scaly and shining white like the cap. This pure-looking, handsome mushroom is one of the most poisonous of its kind. It is called Amanita virosa—the poisonous Amanita, from a Latin word meaning poison. We have never found any specimen with insects on it. They seem to know its deadly qualities and shun its acquaintance.

Let us look at the gray mushroom and see how it differs from the others. It has no ring, its color is a soft gray or mouse color, the margin is deeply grooved. The cap is almost flat, the flesh does not reach to the margin, and is white. It is very smooth, but another time we might find the same mushroom with scales upon it. The cap measures 3 inches across. The stem tapers upward, is slender, and is 4 inches long. The gills are free, not attached to the stem, and are swollen in the middle. They are not very close together and are shining white. The base extends deep into the ground, and is sheathed with a membrane that is loose and easily broken off. It is a very common mushroom, and we shall often find it, but it varies in color; it is sometimes umber, often white, and even has a faint yellowish or greenish hue in the centre.

So far we have only looked at Amanitas. They are conspicuous, and the large rings and colors are striking and interesting to the novice; but look at that clay bank that borders on our road, and perhaps we may discover some Boleti. Even a beginner in the study of mushrooms can tell the difference between a boletus and those we have been examining. Here are two or three mushrooms growing together. What is there different about them? We see no ring, no membrane around the base of stem, and what are these tubes beneath the cap so unlike the gills of the others? They have the appearance somewhat of a sponge. These are the pores or tubes that contain the spores. Let us divide the fungus. At the first touch of the knife, through the stem, the color begins to change, and in a moment stem, tubes, and cap turn to a bright blue. We can see the color steal along, at first faintly, and then deepen into a darker blue. The cap is a light brownish yellow color, 2 inches broad, covered with woolly scales. The tubes are free from the stem. They have been white, but are changing to yellow. The mouths or openings of the tubes are becoming bluish-green. The stem is swollen in the middle. It is covered with a bloom. It is stuffed with a pith, and tapers toward the apex. It is like the cap in color, and measures 1½ inch in length. The mouths of the tubes are round. This is Boletus cyanescens, or the bluing Boletus, as named by Professor Peck in his work on Boleti. He says it grows more in the North, and sometimes is much larger than the one we found.

We turn to the bank in hopes of discovering another, and see, instead, what appears to be a mass of jelly half-hidden in the clay, and in the midst some bright scarlet cherries, or at least something that resembles them. We take the trowel and loosen them from the earth, and there, among the gelatinous matter, we find small round balls as large as a common marble, covered by a bright red skin. When cut in half we see they are filled with a pure white substance, like the inside of a young puff-ball. This is quite a discovery. We must look in our books for its name. It is not in our British manual, but we learn from Professor Peck that it is called Calostoma cinnabarinus. Calostoma is a Greek word meaning beautiful mouth, and cinnabarinus is taken from cinnabaris, which means dragon's-blood. We are not responsible for the names given to plants, but cannot help wishing that some might be changed or shortened.

We could go on prolonging our search, and describe many wonderful fungi, so easily found on a summer day, but as our object is to excite curiosity and interest and not fatigue the reader, we will here pause, and afterward arrange the descriptions of mushrooms in a separate section. The ones we have described may be found in the Middle States and in New England.

Chapter 2

MUSHROOMS

ANTIQUITY OF FUNGI

Fungi have existed from early geological ages. They flourished in the Carboniferous period, when the enormous beds of coal were formed, a space of time that occupied many millions of years. Bessey says that the oldest known member of the order of membrane fungi, Hymenomycetes, was called by the name of "Polyporites Bowmanii." During the Tertiary period members of the genera now known under the names of Lenzites, Polyporus, and Hydnum were all in existence. It is interesting to know that even before the Tertiary period the undergrowth consisted of ferns and fleshy fungi. What a time of delight for the botanist! But there were no human beings in those days to roam amongst that luxuriant undergrowth, and only the fossil remains in the deposits of coal and peat are left to tell of their former existence.

MANNER OF GROWTH

Fungi are either solitary, grow in clusters, in groups, or in rings and arcs of circles.

The species called the Fairy mushroom, Marasmius oreades, is the most familiar of all those that grow in rings. Besides this there is the Horse mushroom, Agaricus arvensis; the Chantarelle,

Cantharellus cibarius; the Giant mushroom, Clitocybe maximus, and St. George's mushroom, Tricholoma gambosa. The latter species is reproduced in rings every year. It is a popular saying that when the ring is unbroken there will be a plentiful harvest the following season. It is an early mushroom, appearing in April. It derives its name from the fact of its appearing about April 23d, which is St. George's day in the English calendar. Besides these mushrooms there is another Tricholoma, T. tigrinus, the Tiger mushroom, which sometimes appears in circles. The word tigrinus means a tiger. The cap is variegated with dark brown spots, hence the name. Then there is the Limp Clitocybe, C. flaccida, so called because flaccida means limp. It also appears in rings (according to Stevenson), while the stems are united under the soil.

The waxy Clitocybe, C. laccata, is not spoken of as having that mode of growth in circles, but we have seen many of these mushrooms appearing in arcs of circles, and forming almost perfect rings, particularly after showers of rain, and always on the sides of roads.

Many fairy rings have lasted for years and are very old. We have read of one, in the county of Essex, England, that measured 120 feet across. The grass that covered it was coarse and of a dark green color. What causes these fairy rings? An explanation is given in a newspaper extract from "Knowledge," in which it is said: "A patch of spawn arising from a single spore or a number of spores spreads centrifugally in every direction, and forms a common circular felt, from which the fruit arises at its extreme edge; the soil in the inner part of the disc is exhausted, and the spawn dies or becomes effete there, while it spreads all around in an outward direction and produces another crop whose spawn spreads again. The circle is thus continually enlarged, and extends indefinitely until some cause intervenes to destroy it. The peculiarity of growth first arises from a tendency of certain fungi to assume a circular form."

The perplexing mushroom, Hypholoma perplexum, often grows in clusters, and so does the inky Coprinus, C. atramentarius, also the glistening Coprinus, C. micaceus. The honey-colored mushroom, Armillaria melloea, is often found in crowded clusters, and this growth is common to many fungi.

ODOR

Many mushrooms have distinct odors and are distinguished by this feature. The genus Marasmius may be known by the garlic-like smell peculiar to it, but it never has a mealy perfume. There is one species, the disgusting mushroom, M. impudicus, that Stevenson says has a strong, unpleasant odor; this is also the case in two other species, the ill-odored mushroom, M. fœtidus, and the penetrating mushroom, M. perfurans.

The Chantarelle, Cantharellus cibarius, has the smell of a ripe apricot, a delicious odor and easily detected. One of the Lepiotas, the tufted Lepiota, L. cristata, has a powerful smell of radishes. Some Tricholomas have a strong odor of new meal. The fragrant Clitocybe, C. odora, has the smell of anise.

Coprinus atramentarius
Photographed by C. G. Lloyd.

There is a very small white, scaly mushroom, never more than an inch across the cap, and with a stem hardly two inches high, that has the distinction of possessing the strongest smell of all the membrane fungi (Hymenomycetes). It is called the narcotic Coprinus, C. narcoticus, and it derives its name from its odor. It is very fragile and grows on heaps of manure.

DURATION

There is another Coprinus, the radiating Coprinus, C. radiatus, so called from the radiating folds on the cap, that may carry off the honor of being the shortest-lived of all the membrane fungi. Stevenson says "it withers up with a breath." It is often overlooked, as it perishes after sunrise. It grows in troops, and is perhaps the most tender of all mushrooms.

The genus Marasmius, belonging to the white spored Agarics, has the power of reviving under moisture after withering, so it may represent a genus that endures longest. None of the fleshy fungi have long lives.

USES

Besides the uses of fungi as scavengers of creation, there are some which have a commercial value and yield an article called "amadou." This is a French word, used for a sort of tinder or touch-wood, an inflammable substance which is prepared from a fungus,[1] Boletus igniarius, and grows upon the cherry, ash and other trees. It is made by steeping it in a strong solution of saltpetre and cutting it in small pieces. It is also called German tinder. Thomé says that Boletus laricis and Polyporus fomentarius yield the "amadou" of commerce. Then, again, the birch Polyporus, P. betulinus, is used for razor strops. We need not say anything on the uses of fungi as articles of food. This subject has been exhausted by many able mycologists,

---

1. Worcester's Dictionary, citing Brande.

and, excepting the mere mention of some mushrooms that are edible, the authors have abstained from this part of the subject.

HABITAT

It is interesting to observe where different mushrooms love to dwell. Some are always found on roadsides, as if seeking the notice of passers-by. These are the Clitocybes and Stropharia, and many of the cup-fungi, while the Boleti take shelter in clay banks and hide in every cranny and nook that they can find. Russulas are seen in open woods, rising out of the earth, also the Lactarius, which seems to like the shade of trees. The Cortinarius also prefers their shelter. The Coprinus loves the pastures and fields, near houses and barns, and dwells in groups upon the lawns. The Hypholoma grows in clusters on the stumps of trees. Marasmius is found among dead twigs and leaves. The white Amanitas flourish in woods and open ground. There are some, like Pleurotus, that grow in trunks of trees, and make their way through openings in the bark. Every dead tree or branch in the forest is crowded with all species of Polyporus, while carpets, damp cellars, plaster walls and sawdust are favorite abodes of many fungi.

STRUCTURE AND GROWTH

Mushrooms consist wholly of cells. These cells do not contain either starch or the green coloring-matter, called chlorophyll, which exists in other plants. They are either parasites or scavengers, and sometimes both. The food of fungi must form a part of some animal or plant. When they commence to grow it is by the division of cells, not laterally, but in one direction, upward. As the mushroom grows the stem lengthens, the cap expands and bursts the veil that surrounds it, and gradually gains its perfect shape.

Every mushroom has a spore-bearing layer of cells, which is called the hymenium. This hymenium is composed of a number of swollen, club-shaped cells, called basidia, and close to them, side by side, are sterile, elongated cells, named paraphyses. In the family called Hymenomycetes there are mixed with these, and closely packed together, one-celled sterile structures named cystidia.

The basidia are called mother-cells because they produce the spores.

There is one great group of fungi called Basidiomycetes, so named from having their stalked spores produced on basidia.

The basidia are formed on the end of threadlike branched bodies which grow at the apex, and are called hyphæ. On top of the basidia are minute stalk-like branches, called sterigmata (singular sterigma), and each branch carries a naked spore. They are usually four in number. This group of Basidiomycetes is divided into (1) Stomach fungi (Gasteromycetes), (2) Spore sac fungi (Ascomycetes), and (3) Membrane fungi (Hymenomycetes).

Mycelium

The Mycelium is commonly called the spawn of mushrooms.

It is the vegetative part of the fungus, and is composed of minute, cylindrical, thread-like branching bodies called hyphæ. When we wish to cultivate mushrooms we plant the spawn not the spores. The thread-like branches permeate the earth or whatever the mushroom grows upon. The color of the mycelium is generally white, but it may also be yellow or red. Its structural details are only visible through a microscope.

Every fungus does not bear the spores exposed upon the cap nor underneath it. The first group of Gasteromycetes, or "Stomach fungi," as Professor Peck has called them in his work on "Mushrooms and Their Uses," have the spore-bearing surface enclosed in a sac-like envelope in the interior of the plant. The genus Lycoperdon belongs to this group, and it contains the puff-balls so common in this country.

In the second group, Ascomycetes, or "Spore sac fungi," the spores are produced in delicate sacs called asci. The fruit-bearing part is often cup-shaped, disc-like, or club-shaped, thicker at the top or covered with irregular swellings and depressions like the human brain.

The Morels and Helvellas belong to this group. One often meets with mushrooms of the former genus in the spring, and they are striking and interesting looking fungi. There are many of both genera that are edible. They will be described in detail later.

Botanists have classified Agarics by means of the color of the spores, and it is the only sure way of determining to what class they belong. We propose in this work also to enumerate the mushrooms according to the color of the pileus or cap, and give a list, with a description of each, after this arrangement. This, of course, is merely superficial, but may interest and attract a beginner in the study of fungi. This list will be placed at the end of the book.

The descriptions will be preceded by a classification according to color of spores, some hints to students, and aids to learning which have been found useful to others.

It is appalling to a beginner when he first reads the long list of names of classes, genera, and species, as the latter are so closely allied in resemblance. One has not always the time nor inclination to condense facts for himself, nor to collect necessary information so as to remember it most easily, all which has to be done in the absence of an American manual or textbook. A great deal has been written for us, it is true, by experienced botanists, but a general and comprehensive work has yet to be compiled.

Before we begin our list of fungi, let us learn what a mushroom is, and know something of its component parts. A mushroom consists of a stem and a cap, or pileus. The cap is the most conspicuous part. The color varies from white and the lightest hues of brown up to the brightest yellow and scarlet. Its size is from an eighth of an inch to sixteen inches and more in diameter. The surface is smooth or covered with little grains (granular) or with minute scales (squamulose) shining like satin, or kid-like in its texture. It may be rounded and depressed (concave), elevated (convex), level (plane), or with a little mound in the centre (umbonate). It may be covered with warts, marked with lines (striate), or zoned with circles. The margin may be acute or obtuse, rolled backward or upward (revolute), or rolled inward (involute); it may be thick or thin.

The Stem

The stem is the stalk that supports the cap. It is sometimes attached to one side, and then it is said to be lateral or between the centre and side, and it is called eccentric; when it is in the middle, or nearly so, it is central.

It is either solid, fleshy, stuffed with pith, or hollow, fibrous, firm and tough (cartilaginous). It is often brittle and breaks easily, or it will not divide evenly in breaking. Its color and size both vary, like the cap. It may taper toward the base, or toward the apex, be even or cylindrical. Its surface may be smooth (glabrous), covered with scales (squamulose), rough (scabrous), dotted, lacerated, or be marked with a network of veins (reticulated). The base may be bulbous, or only swollen (incrassated), and it may root in the ground.

Gills free

Gills adnexed

Gills adnate

Gills decurrent

Gills sinuous

Gills serrated

Pileus umbonate

Pileus umbilicate

Margin involute

Margin revolute

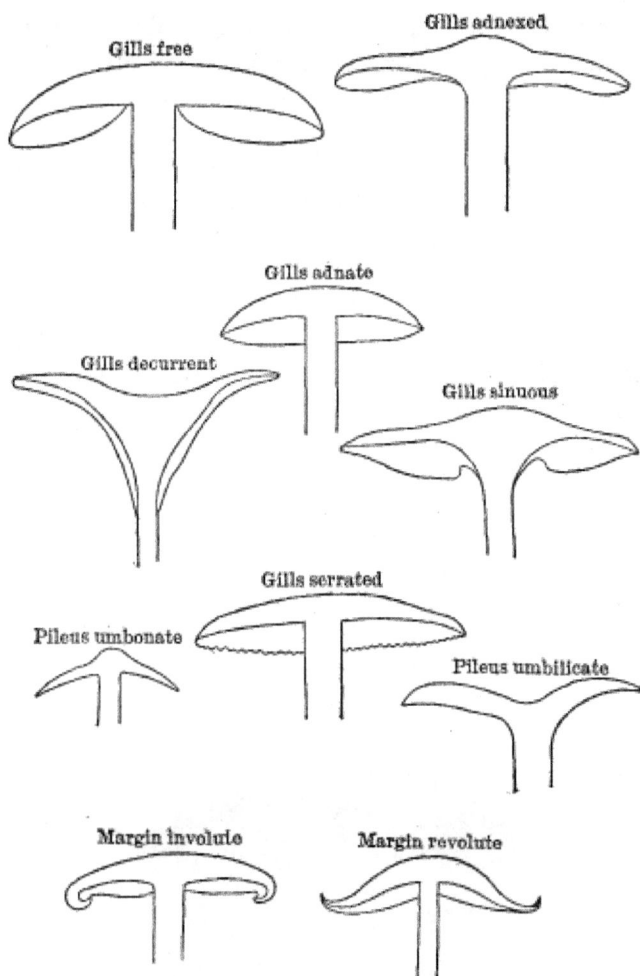

Sections of gill bearing mushrooms.

The Gills

The gills or lamellæ are the radiating parts, like knife blades, that extend from the centre to the margin underneath the cap. They contain the spores. The group of mushrooms that have gills are called Agaracini or Agarics. The gills vary in color; sometimes they change color when mature. When they are close together they are called crowded, and when far apart distant. There are often smaller gills between the others, and sometimes they are two-forked (bifurcate), and are connected by veins.

They are narrow or wide, swell out in the middle (ventricose), are curved like a bow (arcuate), and have a sudden wave or sinus in the edge near the stem (sinuate).

There are various modes of attachment to the stem. Where the gills are not attached to it they are called free; slightly so, adnexed; and when wholly fastened they are adnate. They may run down on the stem, and are then called decurrent.

Amanita vaginata
(breaking from volva)
Photographed by C. G. Lloyd

The Spores

The color of the spores can be seen by cutting off the cap, and laying it gills downward, on a sheet of paper, two or three hours or more. The impression will remain on the paper. It is better to use blue paper, so that the white spores can be seen more clearly. The Agarics are divided into classes according to the color of the spores, so it is of great importance to examine them. The shape and size of the spores can only be learned by the use of a microscope. We have not attempted in this elementary work to do more than mention them.

The Volva and Veil

The universal veil or volva is a thin covering which encloses the entire young plant. The cap grows and expands and bursts this veil into fragments. That part of the veil which breaks away from the cap, called the secondary veil, forms the annulus or ring. It resembles a collar, and is generally fastened to the stem. It is not always permanent or fixed in one place. It may disappear when the plant is mature. It is often fragile, loose and torn, and sometimes is movable on the stem.

The name volva is particularly given to that part of the universal veil which remains around the base of the stem, either sheathing it or appressed closely to it, or in torn fragments. The volva and ring, or annulus, are not always present in mushrooms. The rupture of the veil often causes a part of it to remain on the cap in the shape of warts or scales. These may disappear as the plant grows older, and are sometimes washed off by a heavy rain.

The Tubes or Pores

There is a group of fungi called Polyporei, which have tubes or pores instead of gills. They are placed under the pileus just as the gills are situated, and contain the spores. The length of the tubes varies. The mouths or openings are also of different shapes and sizes. They are sometimes round, and at other times irregular. The color of the mouths is often different from the tubes, and changes when mature. The mouths, too, are sometimes stuffed when young. The attachment to the pileus is to be noted. They may be free or easily detached, depressed around the stem or fastened to it (adnate.)

CLASSIFICATION OF FUNGI

The color of both gills and tubes is an important feature in the classification of fungi.

We have now arrived at a point where the amateur may become wearied at the reading of long names and the enumeration of classes and genera. Stevenson has said in his preface to his work on British Fungi that "there is no royal road to the knowledge of fungi," and if we become enough interested to pursue the subject we will probably discover it at this point. We will try and make this part as simple as possible, and only mention those genera which are most common.

Mushrooms may be divided into three great classes:

i. Gasteromycetes, or "Stomach fungi," where the spores are produced within the plant.

ii. Ascomycetes, or "Spore sac fungi," where the spores are produced in delicate sacs called asci.

iii. Hymenomycetes, or "Membrane fungi," where the spores are produced on the lower surface of the cap.

Class III Hymenomycetes, or Membrane Fungi

This class is divided into six orders:

1. Gill-bearing mushrooms, Agarics, or Agaricini.
2. Fungi with pores or tubes, Polyporei.
3. Fungi with awl-shaped teeth or spines, Hydnei.
4. Fungi with an even spore-bearing or slightly wrinkled surface, Thelephorei.

5. Plants, club-shaped and simple, or bush-like and branched, Clavariei.

6. Gelatinous plants, irregularly expanded, Tremellinei.

The first order, the Agarics, contains most of the well-known mushrooms, as well as most of the edible ones. They have been divided into different classes according to the color of the spores. In a great many cases the color is the same as that of the gills; but this is not always the case, especially in the young plants. The Agarics are divided into four sections:

1. White spores, Leucosporæ.

2. Rosy, salmon or pinkish spores, Rhodosporæ.

3. Brown or ochraceous spores, Ochrosporæ.

4. Dark purplish or black spores, Melanosporæ.

There are an infinite number of mushrooms we shall not mention. The study of fungi has only begun in this country, and there is an immense vista for future students. The amateur or beginner may be well satisfied if after one summer spent in studying mushrooms he can remember the distinguishing types of the various genera, and can say with certainty, "This is a Russula, or this a Cortinarius, or this a Tricholoma." He will then feel he has taken one important step in this "royal road."

DISTINCTIVE CHARACTERISTICS OF GENERA

Hymenomycetes

Order 1 *Agarics*

The names of the genera are all derived from Greek and Latin words. Stevenson, in his book on British Fungi, has given the original words and also their meanings. We take the liberty of copying the English term only, and will place it beside the name of each genus.

Section 1 White Spores, or Leucosporae

The first genus we will mention is:

HYGROPHORUS, from a word meaning moist

This genus contains plants growing on the ground. They soon decay. The cap is sticky or watery, the gills often branched. It has a peculiarity in the fact that the hymenial cells, or the layer of mother cells, contained in the gills, change into a waxy mass, at length removable from the trama. The trama is that substance which extends with and is like in structure to the layer of mother cells.[1] It lies between the two layers of gills in Agarics. The gills seem full of watery juice, and they are more or less decurrent, i. e., extend down the stem. This genus contains many bright-colored and shining species.

We are obliged to refer to the hymenial layer in this place, though the beginner will scarcely understand the meaning of the term. The distinguishing peculiarity of this genus consists in the cells changing to a waxy mass. In the chapter on the structure of mushrooms we have tried to explain something about the cells and the Hymenium.

LACTARIUS = milk

This genus is fleshy, growing on the ground; the cap is often depressed in the centre. The gills are adnato-decurrent, that is, partly attached and prolonged down the stem. They are waxy, rather rigid and acute at the edge. The distinctive feature is the milk that flows when the gills are cut. Sometimes the milk changes color.

---

1. In the young plant it forms the framework of the gills.

RUSSULA = red

This genus grows on the ground, is fleshy, and soon decays. The cap is depressed, or becomes so at a later stage of growth. The stem is polished, generally white, and is very brittle. The gills are rigid, fragile, with an acute edge, and mostly equal in length. Some species exude watery drops. It contains many species of beautiful colors.

CANTHARELLUS = vase or cup

The principal characteristic of this genus consists in the fold-like nature of its gills. The gills are thick, with an obtuse edge, and are branched and decurrent. The genus is fleshy, soft, and putrescent, and has no veil. Some plants grow on the ground and others on mosses.

MARASMIUS = to wither

The genus is tough and dry, not decaying, but shrivelling, and reviving when wet. The stem is tough (cartilaginous.) The gills are rather distant, the edge acute and entire. The plants often have a peculiar smell and taste, like garlic. They are small and thin, commonly growing on the outside of another plant (epiphytal) on the ground, on putrid leaves, or on roots of grasses.

AMANITA

The origin of this name is doubtful. Galen, an ancient Greek physician, is said to have given the name to some edible fungi (Stevenson). It is distinguished as the only genus that hasboth volva and ring. The young plant is enveloped by a universal veil which bursts at maturity. The volva around the base of the stem is formed by the splitting or bursting of the veil, and its different modes of rupture mark the several species. It is sometimes shaped very prettily, and has the appearance of a cup around the stem. It contains many poisonous as well as edible mushrooms.

LEPIOTA = a scale

This genus has a universal veil. The gills are free. Sometimes the ring, or annulus, is movable on the stem. The cap is often covered with warts, or the skin torn into scales, and the stem sometimes inserted in a cup or socket.

ARMILLARIA = ring or bracelet

There is no universal veil in this genus, only a partial one that forms a ring, or sometimes only indicating the ring by scales. The species usually grow on the ground.

TRICHOLOMA = from two Greek words, hair and fringe

This genus is especially noted for its sinuate gills. They have a tooth next to the stem. All grow on the ground and are fleshy. There are sometimes fibrils which adhere to the margin of the cap, the remains of the veil. There are no plants in this genus that are considered poisonous.

CLITOCYBE = a declivity

The gills in this genus are attenuated behind and are attached to stem (adnate) or run down it (decurrent.) The cap is generally plano depressed or funnel-shaped (infundibuliform). Some are fragrant; the odor resembles fresh apricots.

COLLYBIA = a small coin

The stem in this genus is tough or stuffed with a pith, and covered with a cartilaginous rind. The margin of the cap is smooth and turned under at first (involute). The gills are soft, free, or only adnexed behind. The plants grow on the outside of wood and leaves, even on fungi, but are often rooted on the ground, and do not dry up. The gills are sometimes brightly colored.

MYCENA = a fungus

In this genus also the stem is cartilaginous, the cap is sometimes bell-shaped (campanulate) and slender. The plants are generally small and fragile. The cap is from ⅛ to 1½ inch broad. The stem is sometimes filiform, and they grow on stumps and sticks, dead wood, twigs and leaves. They may be found early in the season, but oftener from August to November.

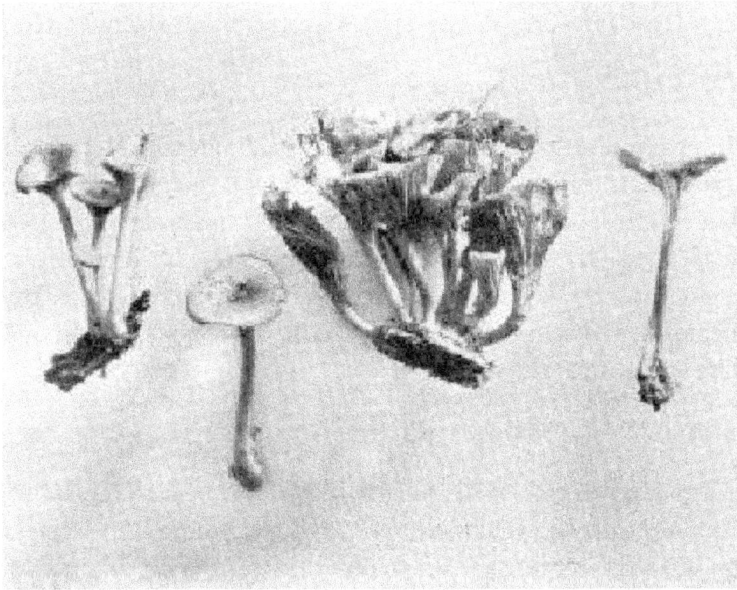

Omphalia alboflava
Photographed by C. G. Lloyd.

OMPHALIA = depressed

The stem in this genus is cartilaginous. The gills run down the stem. The cap is somewhat membranaceous. It is oftener depressed and funnel-shaped. The gills are often branched. The species grow in moist places. The plants are generally small. The largest only measure 2 inches, the smallest only ½ inch across the cap.

PLEUROTUS = side and an ear

In this genus the stem is sometimes wanting, or it grows on the side, or between the centre and margin (eccentric). The plants rarely grow on the ground. They are irregular and fleshy or membranaceous. The time of growth is generally in the autumn. There are a few edible species.

Section 2 Rhodosporae, Red or Pink Spores.

In this section of Agarics the spores are red, pink, or salmon color.

PLUTEUS = a penthouse

This genus has neither volva nor ring. The gills are rounded behind and free, entirely separate from stem, white, then flesh-colored, but often tinged with yellow. The cuticle is sometimes covered with fibres, or with a bloom upon it (pruinose). The apex of the stem is inserted in the cap like a peg, and in this it resembles the Lepiotas. The species grow on or near trunks, appear early, and last until late in the season.

ENTOLOMA = within and fringe

This genus resembles Tricholoma, which belongs to the white-spored Agarics and Hebeloma, which is rosy-spored. The species grow on the ground, and are found chiefly after rain. The stem is fleshy or fibrous, soft, sometimes waxy. The cap has the margin incurved, the gills have a tooth (sinuate), and are adnexed to the stem. Some species smell of fresh meal.

Section 3 Ochrosporae, Brown or Ochraceous Spores

CORTINARIUS = a veil

This genus has a veil resembling a cobweb. The gills generally become cinnamon colored. They grow on the ground in woods, during late summer and autumn. Some of our most beautiful mushrooms belong to this group. The veil is not persistent, and soon disappears.

PHOLIOTA = a scale

This genus mostly grows on trunks. The partial or secondary veil takes the form of a ring. The cap is often covered with scales.

INOCYBE = fibre and head

This genus is distinguished by the silky fibrilose covering of the cap, which never has a distinct pellicle, and by the veil which is lasting and of like nature to the fibrils of the cap. All grow upon the ground.

HEBELOMA = youth and fringe

In this genus the margin of the cap is at first incurved. The gills are attached with a tooth, with the edge more or less of a different color, often whitish. The stem is fleshy, fibrous, somewhat mealy at the apex. They grow on the ground and are strong-smelling, appear early in the autumn, and continue until late in the season.

PAXILLUS = a small stake

This genus is fleshy, putrescent; at first the cap has the margin turned under (involute), then it unfolds gradually and dilates. There are some species of both Tricholoma and Clitocybe that resemble it. The gills separate easily from the cap, and in this it is similar to the Boleti, where the tubes separate also with ease.

Section 4 Melanosporae, Dark Purple or Black Spores

PSALLIOTA = a ring or collar

The common mushroom Agaricus campestris belongs to this group. The gills are rounded behind and free, the stem has a collar. There are many edible mushrooms in this genus. They grow in pastures, and the larger ones are called Champignons. In former times when one spoke of eating mushrooms the species A. campestris, or campester, was always the one denoted.

STROPHARIA = a sword belt

This genus has a ring. The gills are generally attached to the stem; some species grow on the ground, and some grow on other fungi. They are sometimes bell-shaped and then flattened, often with a mound or umbo.

HYPHOLOMA = web and fringe

The veil in this genus is woven in a web which adheres to the margin of the cap. The cap is more or less fleshy, and the margin at first incurved. The gills are attached or have a tooth. There is no ring. The plants grow in tufts on wood, or at the base of trees in the autumn.

PSILOCYBE = naked and head

The cap in this genus is fleshy, smooth, and the margin at first incurved. Gills turn dusky purple. The stem is cartilaginous, hollow or stuffed. No veil is visible. They grow on the ground.

PSATHYRA = friable

The cap is conical and soft, the margin at first straight, and then pressed to the stem. The plants are slender, fragile and moist. Gills become purple. They grow on the ground, or on trunks of trees.

COPRINUS = dung

In this genus the spores are black. It has two distinctive features: one, that the gills cohere at first, and are not separated when young; and the other, that they dissolve into an inky fluid. The gills are also scissile, that is, they can be split, and are linear and swollen in the middle. The plants last but a short time. Some are edible.

Order 2 Polyporei, or Tube-Bearing Fungi

We now pass to the next order, the Polyporei. We will mention four genera:

BOLETUS

The name is that of a fungus much prized for its delicacy by the Romans, and is derived from a Greek word meaning a clod, which denotes the round figure of the plant.

The Boleti grow on the ground, are fleshy and putrescent with central stems. The tubes are packed closely together and are easily separated.

FISTULINA = a pipe

In this genus the tubes are free and distinct from one another. They are somewhat fleshy and grow upon wood.

POLYPORUS = many pores

The pores or tubes in this genus are not separate from one another. They are persistent fungi, most of them growing upon wood.

DAEDALEA = curiously wrought

The name of this genus is derived from Daedalus, who constructed the labyrinth at Crete, in which the monster Minotaur was kept. It was one of the seven wonders of the world.

These fungi grow on wood, and become hard. The pores are firm when fully grown; they are sinuous and labyrinthine.

Order 3 *Hydnei, or Spine-Bearing Fungi*

The name is derived from a word meaning a spine. This order contains many genera, two of which we will mention, Hydnum and Tremellodon.

HYDNUM

Hydnum is derived from a Greek word, the name of an edible fungus. The plants in this genus are furnished with spines or teeth, instead of gills or tubes, and these contain the spores. The species are divided according to the stem. In some it is central and grows on the ground, in others it is lateral, and the cap is semicircular (dimidiate), and others again have no stem. There are some species that have no cap, and the spines are either straight or oblique. There are a few that are edible, but generally they have a bitter taste. However, some writers say that Hydnum repandum, or the spreading Hedgehog, is "delicious." This mushroom and the one named "Medusa's head," H. caput Medusæ, are perhaps the most conspicuous of the order. The latter is very large. Its color is at first white, then becoming ashy gray. The spines on the upper surface are twisted, while the lower ones are long and straight. It grows on trunks of trees. In the spreading Hydnum the margin of the cap is arched and irregular. It grows on the ground.

TREMELLODON = jelly and a tooth

The fungi in this genus are gelatinous. The cap is nearly semicircular in shape, sometimes fan-shaped and rounded in front. The spines or teeth are soft, white and delicate. We found one specimen in the month of September in the mountains of the State of New York.

Order 4 *Thelephorei, or Even Surface Fungi*

In this order the lower surface of the cap is smooth and even, or slightly wrinkled. It is divided into several genera, only two of which we will enumerate, Craterellus and Stereum.

CRATERELLUS = a bowl

The species called the "horn of plenty," Craterellus cornucopioides, belongs to this genus, and is often found. Stevenson says it is common. It is trumpet-shaped (tubiform). The cap is of a dingy black color, and the stem is hollow, smooth, and black. We found quite a small specimen, the pileus not more than 1½ inch broad, but it may measure 3 inches. The spore-bearing surface was of an ash color. The margin of the cap was wavy, and it was hollow right through to the base. It was only 2 inches high, and there was scarcely any stem.

STEREUM = hard

The genus Stereum is woody and leathery in nature, somewhat zoned, and looks like some Polyporci. It grows on wood, on stumps, and on dead wood.

Order 5 Clavariei, or Club Fungi

This order contains several genera, but one only will be mentioned, that of Clavaria.

CLAVARIA = club

The common name often given to this genus is "Fairy Clubs." We have described several species in our list of fungi, and will only say that these are fleshy fungi, either simple or branched. The expression fleshy, so often met with in these pages, is used in speaking of plants when they are succulent and composed of juicy, cellular tissue. They do not become leathery. In the genus Clavaria the fungi have no caps, but they have stems. There are a few edible species. One can scarcely walk any distance without seeing some species of Clavaria. They are conspicuous, sometimes attractive looking, and interesting in their variety.

The genus Cortinarius, one of the order of Agarics, has been already described, but it contains so many species that it deserves especial mention.

They are difficult to define. The genus has been subdivided by botanists into tribes which it may be well to enumerate. We have followed Stevenson's arrangement.

He divides Cortinarius into six tribes.

1. Phlegacium = clammy moisture. In this tribe the cap is fleshy and sticky (viscous), while the stem is firm and dry. In all Cortinarii the gills become cinnamon-colored. There are many large-sized mushrooms in this tribe, the cap sometimes measuring 6 inches across.

2. Myxacium = mucous. This tribe has the stem sticky (viscous), and the universal veil is glutinous. The cap is fleshy but thin. Gills attached to stem and decurrent.

3. Inoloma = fibre and fringe. It contains distinguished species. The cap is at first silky, with innate scales or fibrils, is equally fleshy and dry. The stem is fleshy and rather bulbous.

4. Dermocybe = skin and head. The cap and stem are both thinner in this tribe than in Inoloma. The pileus becomes thin when old, and is dry, not moist. It is at first silky. The color of the gills is changeable, which makes it hard to distinguish the species.

5. Telamonia = lint. Pileus moist; at first smooth or sprinkled with superficial whitish fibres of the veil. Flesh thin, or becoming so abruptly at the margin; the veil is somewhat double, which is a distinguishing characteristic of this tribe.

6. Hygrocybe = moist and head. Cap in this tribe is smooth or only covered with white superficial fibrils, not gluey, but moist when fresh, and changing color when dry. Flesh thin.

Class I Gasteromycetes, or Stomach Fungi

The Basidia-bearing fungi, or Basidiomycetes, are divided into three classes, as has been already stated. The third class, Hymenomycetes, or Membrane fungi, has been described, but there remain two other groups of which we will now speak more fully. They may be considered too difficult for beginners, and we would not venture to enter further into the subject were it not that some of the most familiar fungi belong to these classes— such as Puff-balls, Morels, and Helvellas.

The first class, called the Gasteromycetes, or Stomach fungi, matures its spores on the inside of the plant. The distinction between this class and that of the Membrane fungi, which ripens its spores on the outside, may be more readily understood by one familiar with the structure of the fig, whose flowers are situated on the interior of its pear-shaped, hollow axis, which is the fruit.

We will divide the Stomach fungi into four orders—1) the thick-skinned fungi (Sclerodermae); 2) the Bird's-nest fungi (Nidulariæ); 3) the Puff-balls (Lycoperdons); 4) the Stink horns (Phalloidæ.)

Order 1 Sclerodermae, The Thick-Skinned Fungi

Our attention will be confined to only one genus, and, indeed, one species of this family. We often see in our walks what at a first glance look like potatoes lying along the road, and the suggestion arises that some careless boy has been losing potatoes from his basket on his way home from the country store. We stoop to pick them up, and find them rooted to the ground and covered with warts and scales. We cut them open and find them a purplish-black color inside. It is a mass of closely packed unripe spores. In a few days the upper part of the outside covering decays, bursts open, and the ripe spores escape. This is called the common hard-rind fungus, or Scleroderma vulgare.

Order 2 Nidulariae, The Bird's-Nest Fungi

This is again divided into three genera. The Crucible (crucibulum), the Cup (Cyathus), the Bird's-nest proper (Nidularia.)

We often find on a wood-pile or a fallen tree some of the members of the Bird's-nest family. It is fascinating to examine them in their various stages of development. First we see a tiny buff knot, cottony in texture and closely covered; next, another rather larger, with its upper covering thrown aside, displaying the tiny eggs, which prompts one to look around for the miniature mother bird; then we find a nest empty with the fledglings flown. The characteristic that distinguishes the Bird's-nest fungi from others consists in the fact that the spores are produced in small envelopes that do not split open, and which are enclosed in a common covering, called the peridium. One species is known by the fluted inside of the covering, which is quite beautiful. They are all small and grow in groups.

Order 3 Lycoperdons, The Puff-Balls

The Lycoperdons contain several genera, among which we select the Puff-balls proper and the Earth stars.

What child is there who lives in the country and does not know the Puff-ball? With what gusto he presses it and watches what he calls the smoke pouring from the chimney. Indeed, the outpouring of myriads of spores in its ripe stage does suggest smoke from a chimney. The puff-ball, when young, is of a firm texture, nearly round, grayish, or brownish outside, but of a pure white within. There are several genera, but we have selected two—1) Lycoperdon; and 2) Earth Star, or Geaster.

LYCOPERDON = the puff-ball

The puff-balls vary greatly in size, the smallest measure ½ inch up to the largest, about 15 inches. Professor Peck describes them thus: "Specimens of medium size are 8 to 12 inches in diameter.

The largest in the State Museum is about 15 inches in the dry state. When fresh it was probably 20 inches or more. The color is whitish, afterward yellowish or brownish. The largest size was called the Giant Puff-ball (Calvatia bovista)."

GEASTER = the earth star

These vary greatly in size. The small ones grow on pine needles on the ground or among leaves. Some are mounted on pedicels, some are sessile or seated directly on the earth, but the family likeness is so pronounced that even the novice need not be doubtful as to the name of the fungus when found. There are two species that have slender, elongated stems. The name is well chosen. In moist weather the points expand and roll back or lie flat on the earth. Then the round puff-ball in the centre is plainly seen.

In dry weather the star-like divisions are rigidly turned in and cover closely the round portion. "When dry it is sometimes rolled about by the wind; when it is wet by the rain or abundant dew it absorbs the moisture and spreads itself out, and rests from its journey, again to take up its endless wandering as sun and rain appear to reduce it once more to a ball and set it rolling." (Underwood.)

Order 4 Phalloids, The Stink Horn Fungi

We come now to the fourth and last order of the Stomach fungi (Gasteromycetes) that we shall mention. In spite of their appellation these fungi are strikingly beautiful, but their odor is most offensive. They grow in woods, and are also found in cellars. Their history has been carefully investigated by mycologists, and the novice will find many beautiful illustrations in various works. In their early stage they are enclosed in an egg-shaped veil (volva), having a gelatinous inner layer. Some are bright-colored, others are pure white, and the stems of one species look as if covered with lace work. The most familiar one, Phallus impudicus, "the fetid wood witch," we have placed in the list of fungi at the end of this book, with its description.

Class II Ascomycetes, or Spore Sac Fungi

This is the second division of the Basidia-bearing fungi. It includes all the fungi that have the spores enveloped in delicate sacs called asci. It is divided into several orders, but we will only mention the one which contains the most familiar plants. This order is named the Disc-like fungi (Discomycetes). In this the spore-bearing surface is on the upper or outside surface of the mushroom cap. It is divided into many genera, of which we shall mention three—the Cup fungi, or Pezizas, the Morels or Morchellas, and the Yellowish fungi or Helvellas.

PEZIZAS = the Cup fungi

These form a very large group, mostly growing on decaying plants. They are typically disc-shaped or cup-shaped, and when young are closed or nearly so, opening when mature. They vary in size from minute species to large fleshy ones, 3 to 4 inches in diameter. They are generally small, thin, and tough. They grow on twigs, leaves, dead wood, or on the ground. Many are stemless. They are both solitary and densely clustered. The color varies from pale brown to a dark gray, resembling, when moist, india-rubber cloth, and then, again, there are many of brilliant hues—red and orange. Some are erect, some are split down at the side like the ear of a hare. The Cup fungi are found in August and September, growing near ditches, and by the roadside where there is moisture. The ear-shaped Pezizas somewhat resemble the Jew's ear, and the beginner might easily confound them. This latter fungus belongs to the third class of membrane fungi (Hymenomycetes), and it is included in the descriptions of fungi.

THE MORELS or MORCHELLAS = the honey-combed fungi

The collector during the months of April and May will enjoy a new experience when he first finds a fungus of a bright brown color, deeply pitted, spongy looking, cone-shaped or nearly round; its head supported on an erect, white stem. He will probably find it on a grassy hillside or along a running brook under some forest trees. He has perhaps seen its picture and at once exclaims, "my first Morel." He will notice its peculiar honey-combed depression, and then cutting it open will find both the head and the stem hollow. Where are the spores? There are no gills as in the Agarics, nor are they concealed in a covering (peridium), as in the Puff-balls, but they are contained in delicate sacs on the cap. The exterior surface of the cap is the spore-bearing portion, and the spores are developed in their sacs, but only seen under a microscope.

HELVELLA = the yellowish mushroom

This genus may be readily recognized by the form of the cap, which is lobed and irregularly waved and drooping, often attached to the stem. They grow on the ground in the woods, and sometimes on rotten wood. The genus comprises the largest of the Disc fungi known, some species weighing over a pound. Cicero mentions the Helvellas as a favorite dish of the Romans.

THE TRUFFLE = delicacy

It will be well to finish this section with the mention of the Truffle. It may yet be found in the United States, but hitherto its place of growth has been on the continent of Europe, and especially in France, where it forms an article of commerce, and is highly prized as food. It is subterranean, and requires for its discovery a higher sense of smell than man possesses. It is generally found by the hog and the dog, who are trained to help the truffle hunters. There are some species in our country that resemble it, and grow underneath the ground. One, found in the Southern States, called Rhizopogon, grows in sandy soil. This species, however, does not belong to Class II., but to Class I., the Gasteromycetes, or Stomach fungi. It is not likely that the beginner will find this mushroom, so no description will be given.

Chapter 3

GENERAL HELPS TO THE MEMORY

There are certain facts which if committed to memory will be of great help to beginners in classifying mushrooms. There are distinctive features belonging to different genera, which will be enumerated as follows. These facts apply to the order of Agarics, containing the largest number of familiar mushrooms. They have been placed in tables for the convenience of the beginner, and are arranged without regard to family relationship.

Mushrooms Containing both Volva and Ring (Annulus).

There is only one genus that has both volva and ring. Amanita.

Mushrooms with Ring and no Volva.

1. Pholiota.

2. Annularia.

3. Stropharia.

4. Psalliota.

5. Armillaria.

6. Lepiota.

Mushrooms that have the stem attached on the side (lateral) or between Margin and Centre (eccentric).

1. Crepidotus.

2. Claudopus.

3. Pleurotus.

Mushrooms with tough or cartilaginous Stems.

1. Psathyra.

2. Nolanea.

3. Mycena.

4. Marasmius.

5. Naucoria.

6. Leptonia.

7. Omphalia.

8. Collybia.

9. Psilocybe.

10. Galera.

Mushrooms, Stemless.

1. Schizophyllum.

2. Trogia.

3. Lenzites.

Mushrooms that have the Cap bell-shaped (campanulate) and Marked with Lines (striate).

1. Psathyra.

2. Galera.

3. Nolanea.

4. Mycena.

Mushrooms with Gills attached to Stem and a Ring.

1. Stropharia.

2. Armillaria.

3. Pholiota.

Mushrooms Having Gills with serrated edge.

1. Lentinus.

Mushrooms with Free Gills not attached to Stem.

1. Chitonia.

2. Psalliota.

3. Pluteolus.

4. Pluteus.

5. Volvaria.

6. Lepiota.

7. Amanita.

Mushrooms with emarginate sinuate Gills, or with notch near to Stem.

1. Hypholoma.

2. Tricholoma.

3. Hebeloma.

4. Entoloma.

Mushrooms that are corky and leathery.

1. Lenzites.

2. Lentinus.

3. Schizophyllum.

4. Panus.

Mushrooms with Gills running down Stem more or less (decurrent).

1. Gomphidius.

2. Paxillus.

3. Tubaria (some species).

4. Flammula (some adnate).

5. Eccilia (truly decurrent).

6. Clitopilus (somewhat decurrent).

7. Panus (some species decurrent).

8. Lentinus (mostly decurrent).

9. Cantharellus.

10. Hygrophorus (mostly decurrent).

11. Pleurotus (some decurrent).

12. Omphalia (truly decurrent).

13. Clitocybe (decurrent or adnate).

14. Lactarius (decurrent or adnato-decurrent).

Mushrooms that are deliquescent or turn into inky fluid.

1. Coprinus.

2. Bolbitius.

It will also be useful to the beginner to see a list of Agarics classified according to botanists by the color of their spores.

CLASSIFICATION OF AGARICS BY COLOR OF SPORES

1. Leucosporæ (white spores).

2. Rhodosporæ (rosy or salmon spores).

3. Ochrosporæ (ochraceous spores).

4. Melanosporæ (dark purple or black spores).

Leucosporae, or White Spores.

1. Amanita.

2. Lepiota.

3. Armillaria.

4. Tricholoma.

5. Clitocybe.

6. Collybia.

7. Mycena.

8. Omphalia.

9. Pleurotus.

10. Trogia.

11. Hygrophorus.

12. Lactarius.

13. Russula.

14. Cantharellus.

15. Marasmius.

16. Lentinus.

17. Panus.

18. Xerotus.

19. Schizophyllum.

20. Lenzites.

21. Arrhenia (pallid spores).

Rhodosporæ, Rosy or Salmon Spores.

1. Volvaria.

2. Pluteus.

3. Enteloma.

4. Leptonia.

5. Nolanea.

6. Eccilia.

7. Claudopus.

8. Clitopilus.

Ochrosporæ, or Ochraceous Spores.

1. Pholiota.

2. Inocybe.

3. Hebeloma.

4. Flammula.

5. Naucoria.

6. Pluteolus.

7. Galera.

8. Tubaria.

9. Crepidotus.

10. Cortinarius.

11. Acetabularia.

12. Paxillus (spores are ferruginous or dingy white).

13. Bolbitius (ferruginous spores).

Melanosporæ, Dark Purple or Black Spores.

1. Chitonia.

2. Psalliota.

3. Stropharia.

4. Hypholoma.

5. Psilocybe.

6. Psathyra.

7. Panæolus.

8. Psathyrella.

9. Coprinus.

10. Gomphidius.

11. Anellaria.

Having arranged these lists of mushrooms by their different characteristics, and then by the color of the spores, we will give a list of fungi familiar to most persons, classified according to the colors of the cap. The far greater number have been analyzed by the writers, and a full description is given to enable the beginner more easily to identify them.

The reader will notice that in the lists of fungi given above there are certain genera not elsewhere mentioned in this book. He will understand that it is inadvisable in a short primer to allude to all the genera that exist. It was, however, impossible to give a complete table without including them in it.

Russula pectinata
Photographed by C. G. Lloyd

Chapter 4

DESCRIPTIONS OF FUNGI, ARRANGED ACCORDING TO COLOR OF CAP ONLY

MUSHROOMS WITH RED OR PINK COLORED CAP

The genus Russula probably contains the largest number of mushrooms with reddish caps, the word Russula meaning reddish.

RUSSULA EMETICA = a vomit.
The Nauseating Russula.

Cap bright blood red, at first rosy, then blood color, tawny when old, 3 to 4 inches broad, first bell-shaped, then flattened or depressed, polished, margin at length grooved (sulcate), flesh white, reddish under the cuticle. **Stem** 1½ to 3 inches long, ¾ of an inch thick, white or with a reddish hue, spongy, stuffed, stout, elastic when young, fragile when old, even, tapering slightly upward. **Gills** free, broad, rather distant, white.

This is found on the ground among dead leaves, in the woods and open places from July to December. It has a bitter taste, and is said to be poisonous. Those eating it are often affected as if they had taken an emetic. It is easily distinguished by the fact of the flesh turning red immediately under the skin when it is peeled off. There are numerous varieties of it, in one the stem has minute wrinkles running lengthwise. We found it in different localities. The taste was acrid. It was one of the first and the last mushrooms that we gathered. (Poisonous.)

RUSSULA SANGUINEA = blood.
The Blood-colored Russula.

Cap blood red, becoming pale at margin, 2 to 3 inches broad, at first convex, then depressed, and funnel-shaped (infundibuliform), irregularly swollen in the centre, polished, even, margin acute, moist in damp weather. Flesh firm, cheesy, white. **Stem** stout, spongy, stuffed, at first contracted at apex, then equal, slightly marked with lines white or reddish. **Gills** at first fastened to stem and then decurrent, crowded, narrow, connected by veins, fragile, somewhat forked, shining white, afterward turning ochraceous color. The taste is acrid and peppery. It is found in woods from August to September, and is not common. (Poisonous.)

RUSSULA ROSEIPES = rosy stem.
The Rosy Stemmed Russula.

This is a striking-looking mushroom. The colors are pretty, and the tinge of red in the stem adds to its beauty. There are other species of Russula that also have red tints in the stem. **Cap** rosy red, with pink and orange hues, 1 to 2 inches broad, convex, becoming nearly plane or slightly depressed; at first viscid, soon dry, slightly marked with lines on the thin margin, taste mild. **Gills** moderately close, nearly entire, rounded behind and slightly adnexed, swollen in the middle, whitish, becoming yellow. **Stem** 1 to 2 inches long, 3 to 4 lines thick, slightly tapering upward, stuffed or hollow, white, tinged with red. It is distinguished from other species by its mild taste, rosy cap, commonly dry and but slightly striate on margin, its gills changing from white to yellow or slightly ochraceous, and being partially attached to the stem, and its stem being slightly stained with rosy red. It grows in pine and hemlock woods, and is found in July and August. (Edible.)

RUSSULA LEPIDA = neat or elegant.
The Elegant Russula.

Cap at first is a bright red, but becomes a dull reddish-pink, paler at the disc, 3 inches broad, dry, fleshy, convex; then expanded, scarcely depressed, obtuse and polished, afterward cracked (rimose), and with minute scales (squamulose). The margin spreading and rounded, obtuse, not striate. **Stem** about 3 inches long, from 1 to 1½ inch thick, even, solid, white, or rose color. **Gills** rounded behind, rather thick, somewhat crowded, often forked, connected by veins, white, often red at edge. Taste mild. We found our specimen in mixed woods. The stem was only tinged with pink. (Edible.)

LACTARIUS VOLEMUS = a kind of large pear.
(From its shape.)
The Orange Brown Lactarius.

Cap 3 to 5 inches broad, reddish-orange color, becoming pale, compact, rigid, obtuse, with the margin bent inward, depressed, at length marked with lines like a river (rimose). Flesh white, turning brown. **Stem** 2 to 3 inches long, ¾ to 1¼ inch thick, stout, stuffed, then hollow, paler at apex, with a bloom, same color as cap, with lengthwise lines. **Gills** adnato-decurrent, yellowish turning ochraceous, broad, thin, crowded, milk sweet and plentiful. Stevenson says that the taste of this Lactarius is delicious, that it is savory even when raw. It should not be kept too long before cooking, or it will emit a strong, unpleasant odor. It is abundant in chestnut or oak woods from July to September. Our specimen was much wrinkled on the margin. The milk was abundant. (Edible.)

LACTARIUS ICHORATUS = lymph.
The Colorless Lactarius.

The name of this species is given on account of the color of the milk (Stevenson). **Cap** a tawny pinkish-red color, 3 to 4 inches broad, zoned, plano-depressed, margin often wavy, dry, flesh creamy white or pallid. **Stem** 1½ to 3 inches long, thick, solid, afterward spongy, equal, smooth, the same color as the cap, lighter at the apex. **Gills** adnate, slightly decurrent, not crowded, creamy white, turning ochraceous. Milk white, sweet. It has a strong smell. In the specimen we found the stem was slightly marked with lines and the milk plentiful. It is not spoken of as edible.

LACTARIUS MITISSIMUS = mild.

The name only applies to the taste of the milk. (Stevenson.)

Cap a light, bright reddish-orange, golden tawny color, 1 to 4 inches broad, even, then depressed, smooth, sticky when moist, flesh whitish, turning yellow. **Stem** 1 to 4 inches long, thick, stuffed, then hollow, even, smooth, same color as cap. **Gills** slightly running down the stem, rounded at one end, broad, yellowish. Milk mild, then bitterish and plentiful. It is found in pine and mixed woods from August until November. It has a beautiful color, and resembles in that particular L. volemus.

CORTINARIUS ARMILLATUS = a ring or bracelet.
The Zoned Cortinarius.

Cap a tawny reddish-yellow, brick red, 2 to 5 inches broad, fleshy, bell-shaped or almost conical, then convex, dry, smooth, marked with reddish specks, darker toward the centre, flesh white, turning red and narrowing toward the margin. **Stem** 3 to 6 inches long, ½ inch thick, solid, firm, slightly tapering toward the apex, very bulbous at base, same color as cap, stuffed with brown pith inside. There are two or three reddish oblique zones encircling the stem. **Gills** adnate, swollen in the middle, distant, variable, at first pale cinnamon color, and then dark brown. We found them at the end of August in great numbers, sometimes united in tufts (cæspitose) in all stages of growth, the younger ones covered with a cobwebby veil, which is paler in color than the zones. They grow in mixed woods.

CLITOCYBE LACCATA = a resinous substance.
The Waxy Clitocybe.

This species is small in size. Cap is about 1 inch broad, thin, convex and almost plane. Sometimes with a depression (umbilicate). When moist it has a water-soaked look, and becomes pale in drying. When wet it has a peculiar flesh color, but when dry it is a pale yellowish-red hue. **Stem** is long and slender, tough and of same color as cap, 2 lines thick, fibrous, stuffed, often twisted and white, with soft, weak hairs at base (villous). **Gills** are attached to stem with a decurrent tooth, broad, distant, of a peculiar flesh color. We found several varieties. One had gills of a beautiful violet color (Var. amethystina), in another the gills were pale (Var. pallidifolia). (Peck.) A small form with radiating lines extending from near the centre to the margin (Var. striatula), Peck, is an interesting species and often seen. They grow closely together on the sides of roads, in groups, all through the season. Sometimes the cap is very small, ¼ inch across. It often grows in arcs of circles.

CLITOCYBE INFUNDIBULIFORMIS = funnel-shaped.
The Funnel-shaped Clitocybe.

Cap a pale red color, 2 to 3 inches broad, convex when young, then slightly raised in the middle, umbonate, afterward the margin is elevated and the cap becomes funnel-shaped and the margin wavy. Flesh thin and white. **Stem** 1½ to 3 inches long, 2 to 3 lines thick, smooth, paler colored than the cap, tapering upward. **Gills** rather decurrent, arc-shaped, broad, distant, whitish, not yellow, netted with veins. This is also a variable species and grows in woods. It is pretty, and is easily known by its shape.

BOLETUS MURRAYI.
Murray's Boletus.

Cap dark red, 1 to 3 inches broad, granulated, convex, with a slight mound or umbo, margin turned upward, flesh yellow. **Stem** ½ inch long, yellow. Tubes lemon color, angular and round, irregular. The stem in our specimen was granulated like the cap.

BOLETUS CHROMAPES = chrome yellow and foot.
The Chrome-footed Boletus.

Cap tawny red, 2 to 4 inches broad, convex or nearly plane, flesh white. Tubes almost attached (subadnate), depressed around the stem, whitish, turning a pinkish-brown color. **Stem** equal or tapering upward, rough whitish color, with reddish specks upon it, but chrome yellow at the base, both outside and inside, and spongy within. Stem 2 to 4 inches long, about ½ inch thick. This is not a hard boletus to distinguish on account of the yellow color at the base of the stem. The Boleti seem to be most abundant from the beginning of July until early in September. There are many varieties of beautiful colors, and they are a most interesting group, especially to beginners. This may be partly owing to the fact that Professor Peck's pamphlet on "Boleti" is clearly expressed, and the descriptions so vivid and plain that one has less trouble in naming them than any other class of fungi.

HYGROPHORUS MINEATUS = vermilion.
The Vermilion Hygophorus.

Cap 1 inch broad, at first vermilion color and then paler, broad, flattened and then even, depressed in centre by the margin becoming elevated. It is thin and fragile at first, even, smooth, and then scaly. **Stem** from 1 to 2 inches long, slender, 1 line thick, a little paler than the cap, equal, round, somewhat stuffed, smooth, shining. **Gills** attached, seldom decurrent, distant, distinct, yellow color, shaded with red. This species is very fragile. It grows in woods or in open country, on mosses or on dead leaves. It may be cæspitose, or grows singly from July to October.

HYGROPHORUS COCCINEUS = scarlet color.
The Scarlet Hygrophorus.

Cap, first bright scarlet and then changing to a paler hue. One to 2 inches broad and even more, convex, plane, often unequal, obtuse, sticky, and even, smooth, flesh of the same color as cap. **Stem** 2 inches long, 3 to 4 lines thick, hollow, then compressed and rather even, scarlet color like cap, but always yellow at the base. **Gills** wholly attached, decurrent, with a tooth, distant, connected by veins, soft, watery, when full grown, purplish at the base, light yellow in the middle, powdery at the edge, fragile. This species grows in pastures, and is common. It is found from August to November.

HYGROPHORUS PUNICEUS = blood red.
The Blood-red Hygrophorus.

Cap 2 to 4 inches broad, glittering blood scarlet, when older becomes paler, at first bell-shaped, obtuse, commonly spread out or lobed, irregular, even, smooth, sticky. Flesh of the same color as cap, fragile. **Stem** 3 inches long, 1 to 1½ inch thick. Solid when young, at length hollow, very stout, swollen in middle, thinner at both ends, marked with lines and generally scaly at apex; when dry either yellow or same color as the cap, always white at first, and often incurved at the base. **Gills** ascending, swollen in middle, 2 to 4 lines broad, distant, thick, white or light yellow, or yellow, and often reddish at base. This is a very handsome species. It is found in pastures from July to November.

MUSHROOMS WITH YELLOW OR ORANGE COLORED CAP

CANTHARELLUS CIBARIUS = food.
The Chantarelle.

Cap bright orange or egg color, first convex, and then depressed, at length top-shaped and smooth. The margin lobed and turning under (involute). Flesh thick and white. **Stem** 1 to 1½ inch long, thickened upward, solid, fleshy. **Gills** running down the stem, thick, distant, fold-like. Stevenson does not give the size of the cap, but our specimen measured 2 inches in breadth. It had an odor like ripe apricots, and a pleasant taste. It is often tufted in its growth. It is found in woods from July to December. This is a very striking looking mushroom and easily distinguished. It often grows in rings or arcs of circles. (Edible.)

HYPHOLOMA FASCICULARE = a small bundle.
The Tufted Hypholoma.

Cap a beautiful reddish color, like a peach; the disc darker, about 2 inches broad, fleshy, thin, convex, then plane, with a slight mound or umbo, even, smooth, dry; flesh a light yellow. **Stem** variable in length, 2 to 9 inches long, 2 lines thick, hollow, thin, incurved or curved, covered with fibres of same color as cap. **Gills** adnate, very crowded, linear, somewhat liquid when mature (deliquescent), sulphur yellow, and then becoming green, taste bitter. It grows in crowded clusters. It is said to be poisonous.

AMANITA MUSCARIA = a fly.
The Fly Amanita.

Cap at first red, then orange, then becoming pale, about 4 inches broad, convex, and then flat, covered with thick fragments of volva; margin when grown slightly marked with lines; flesh white, yellow under the cuticle. **Stem** white, sometimes yellowish, 2 inches long, torn into scales, at first stuffed, then hollow; the attached base of the volva forms an oval-shaped bulb, which is bordered with concentric scales, that is, having a common centre, as a series of rings one within the other. **Ring** very soft, torn, even, inserted at the apex of the stem, which is often dilated. **Gills** free but reaching the stem, decurrent, in the form of lines, crowded, broader in front, white, rarely becoming yellow. It grows in woods from July to November. This mushroom is easily identified by its orange-colored cap, covered with white warts and *pure white stem and gills.* We found several specimens in the woods, all of a most beautiful striking color. (Poisonous.)

AMANITA FROSTIANA.
Frost's Amanita.

Cap a bright yellow, almost orange color, 1½ inch broad, convex or expanded, covered with warts, but sometimes nearly smooth, the margin marked with lines (striate.) **Gills** white or tinged with yellow, free from the stem. **Stem** 2 to 3 inches long, white or yellowish, stuffed, slender, bearing a slight evanescent ring; bulbous at the base, bulb slightly margined by the volva. We found several specimens growing in mixed woods. It is smaller than A. muscaria, more slender, with a beautiful color.

TRICHOLOMA EQUESTRE = a knight.
The Canary Mushroom, so called from its color.

Cap pale yellow, 3 to 5 inches broad, darker at disc, tinged with a brick red hue, and yellow near margin, convex, then plane, wavy, irregular; flesh white, thick. **Stem** 1 to 2 inches long, and ½ to ⅔ inch thick, generally white, sometimes yellow, stout and solid. **Gills** close, deeply notched near the stem, a beautiful pale yellow color, scarcely adnexed, broad, somewhat swollen in middle. It grows in pine woods and appears in the autumn.

TRICHOLOMA SULPHUREUM = sulphur.
The Sulphury Tricholoma.

Cap dingy sulphur yellow color, ½ to 4 inches broad, at first round with a slight umbo, at length depressed, rather silky, then smooth and even. **Stem** 2 to 4 inches long, 3 to 4 lines thick, stuffed, somewhat equal but often curved, rather smooth, striate, sulphur yellow, of same color as cap. **Gills** adnexed, narrowed behind, rather thick, distant, distinct, brighter than the cap. This is also found in autumn in the woods, and is quite common. It has a strange disagreeable odor.

LACTARIUS DELICIOSUS = delicious.
The Delicious Lactarius.

Cap orange brick color, 2 to 6 inches broad, becoming pale, fleshy, when young depressed in centre, margin turned under (involute), then flat and depressed, or funnel-shaped, with margin unfolded, smooth, zoned, slightly sticky. The zones become faded in the old plants. The flesh is whitish or tinged with yellow. **Stem** a little paler than the cap, with spots of deeper orange, 1 to 4 inches long, ⅓ to ⅔ of an inch thick, stuffed, then hollow, fragile. **Gills** running down the stem (decurrent), orange color, crowded, narrow, becoming pale and green when wounded. The milk is orange color. It grows in pine woods and in wet, mossy swamps. It resembles the orange brown Lactarius in size and shape, but the color is different, so we have placed it in the orange-colored section and L. volemus in the red division of colors.

Lactarius insulsus

Photographed by C. G. Lloyd

STROPHARIA SICCAPES = dry and foot.
The Dry Stropharia.

Stropharia is taken from a Greek word meaning sword belt, referring to its ring (Stevenson). Siccapes is from two words meaning dry and foot. It grows on horse manure. Stevenson does not mention this species. It is described by Mr. Peck in the State reports. **Cap** is a light yellow, darker in the centre, ¼ inch to 1 inch broad, bell-shaped, sticky, shiny when dry, even. **Stem** sometimes 4 inches long, slender, straight, dry, base almost club-shaped. **Ring** scarcely perceptible, but forming a whitish zone, shining, persistent, apex of stem whitish, and slightly striate. **Gills** dark gray, almost blackish, the margin paler, adfixed, thin. We found a great many in one place, of all sizes, from 1 line across cap to 1 inch. In some specimens the ring was wanting, but in others it was apparent.

CANTHARELLUS AURANTIACUS = orange yellow.
The Orange Chanterelle.

This species takes its name from its color. **Cap** is orange yellow, 2 to 3 inches broad, fleshy, soft, depressed, often eccentric, with the stem between centre and margin, and wavy, somewhat tomentose and involute at the margin. **Stem** 2 inches long, stuffed, and then hollow, somewhat incurved and unequal, yellowish. **Gills** decurrent, tense, and straight, repeatedly dividing by pairs from below upward (dichotomous) and crowded, often crisped at base, orange color. This species grows in woods, and is often found there during the months of autumn. Some consider it poisonous.

CANTHARELLUS INFUNDIBULIFORMIS = funnel-shaped. The Funnel-Shaped Chantarelle.

Cap yellow when moist, 1 to 2 inches broad, umbilicate, then funnel-shaped, wrinkled on the surface, at length wavy at margin. **Stem** 2 to 3 inches long, 2 lines thick, hollow (fistulose), a little thickened at the base, even, smooth, always a light yellow. **Gills** decurrent, thick, distant, dichotomous, straight, light yellow; when old, ash color (cinereous.) This is found in the woods from July to October.

BOLETUS HEMICHRYSUS = half and golden. The Half Golden Boletus.

The descriptions of the Boleti are all written after comparing the specimens we found with those described in Professor Peck's work on Boleti. We examined and analyzed all those placed on the list. The descriptions written by Professor Peck are so clear and faithful to nature that it makes the task of calling them by name much easier than any other fungi we have studied. **Cap** bright golden yellow, 1½ to 2½ inches broad, convex plane and depressed, with minute wooly scales (floccose squamulose), and covered with a yellow powder (pulverulent), sometimes with cracks (rimose). Flesh thick and yellow. Tubes decurrent, yellow, becoming brown; mouths large, angular. **Stem** short, about 1 inch long, 3 to 6 lines thick, irregular, narrowing toward the base, sprinkled with a yellowish dust, tinged with red. We found it growing on an old stump, in pine woods, in the month of August.

BOLETUS GRANULATUS = granules.
The Granulated Boletus.

This Boletus varies much in color. In our specimen it was a pinkish-yellow, and covered with yellow spots of a darker shade. We found it in all sizes, from 2 to 4 inches broad. **Cap**was convex, nearly plane, viscid when moist. It became more of a yellow color when it was dry. Flesh pale yellow. The tubes were adnate, short and yellowish. **Stem** 1 to 2 inches long, 4 to 6 lines thick. Some were united in tufts (cæspitose), others were gregarious (in groups) or solitary. They grew on the edge of pine woods, and near the roadside. The stem was dotted in the upper part with glandules and was pale yellow.

BOLETUS CYANESCENS = bright blue.
The Bluing Boletus.

Cap a light pale brownish-yellow, or a light yellow color (alutaceous), 2 to 5 inches broad, with minute wooly scales, convex or nearly plane. Flesh white, changing quickly to blue when cut. Tubes free, white, afterward yellow; mouths small, round. Tubes change also to a bluish-green when bruised. **Stem** 2 to 4 inches long, ¾ to ½ inch thick, swollen in the middle (ventricose), covered with a bloom (pruinose), stuffed and then hollow, tapering toward the apex, colored like the cap. This is a very easy Boletus to distinguish from others, and interesting to the beginner on account of the striking and beautiful change of color. Found in hemlock and pine woods toward the end of August.

PHOLIOTA ADIPOSA = fat.
The Stout Pholiota.

Cap bright yellowish or orange color, 3 to 7 inches broad, convex, then flattened, gibbous, that is, more convex on one side than on the other; viscid, covered with woolly (floccose) scales, which often separate. Flesh whitish. **Stem** 3 to 6 inches long, ½ to 1 inch thick, solid, large at base, first white and then light yellow, with darker scales. **Ring**yellow, and then ironrust color (ferruginous.) **Gills** adnate, slightly rounded, broad at first, yellow and then darker. We were driving through a thick woods when we saw the bright yellow cap of this mushroom peering among the bushes. There was no apparent ring and few scales except on the margin. It was irregularly shaped, fleshy and thick. It was not a typical specimen, and a beginner would have found it difficult to name. The then recent hard rains had washed nearly all the scales from the cap, and the ring was hardly to be seen. It grew on the trunk of a tree in the month of September. Not edible.

PHOLIOTA SPECTABILIS = showy.
The Showy Pholiota.

This Pholiota was found much later in the season. **Cap** is from 2 to 5 inches broad, a golden yellow, then growing paler, fleshy, torn into squamules, dry, flesh thick, hard, sulphur yellow. **Stem** about 3 inches long and 1 inch thick, solid, hard, swollen in the middle, and extending into a spindle-shaped root. It is sometimes smooth and shining and sometimes scaly, sulphur yellow color and mealy *above* the ring. **Gills** adnate, crowded, narrow, at first pure yellow and afterward ironrust color. Gills have sometimes a small decurrent tooth (Stevenson), but our specimen had none. It grew together (cæspitose) on a stump. Not edible.

MARASMIUS OREADES = a mountain nymph.
The Fairy Ring Mushroom.

Cap when young and moist is of a pale yellowish-red, but fades when dry to pale yellow. It is from 1 to 2 inches broad, fleshy, tough, convex, then plane, somewhat umbonate, even, smooth, slightly striate at margin when moist. **Stem** 1 to 2 inches long and less than ¼ inch thick; slender, solid, tough, equal, sometimes cartilaginous, straight, covered with a close woven skin that can be rubbed off. **Gills** free or slightly attached, whitish or creamy yellow, broad, distant, the alternate ones shorter, rounded, or deeply notched at inner end. These mushrooms grow in circles and are called fairy rings. They are found chiefly on lawns and pastures from May till October. We saw one specimen in October. It grew in a waste lot at Kaighn's Point, Camden, N. J. It was solitary, of a brownish-yellow color, the cap 1 inch broad, and the stem 1 inch long. It was growing amidst some ballast plants, the only mushroom there.

COPRINUS MICACEUS = mica.
The Glistening Coprinus.

Cap varies from buff to tawny yellow, 1 to 2 inches broad, bell-shaped (campanulate) or conical (cone-shaped), thin, marked with lengthwise lines, which extend half-way up from the margin. The disc is even and is more highly colored. It is often sprinkled with shiny atoms when young. **Gills** at first whitish, then brown or black. **Stem** 1 to 3 inches long, slender, hollow and white. The spores are dark brown. We found it in great numbers growing on the ground amidst the grass in September and October. It may be seen as early as April. It is a pretty species. (Edible.)

Amanita vaginata
Photographed by C. G. Lloyd

MUSHROOMS WITH GRAY COLORED CAP

AMANITA STROBILIFORMIS = a pine cone.
The Warted Amanita.

Cap light gray, or dingy white when young; 7 to 9 inches broad when expanded fully. It is covered with large pyramidal, persistent warts. The margin is even, and extends beyond the gills. Flesh firm and white. **Stem** 6 to 8 inches long, 1 to 3 inches thick, solid, scaly, tapering upward, with a bulbous base and marked with a series of rings near the root, which extends deep into the ground. **Ring** large, torn. **Gills** white, free, rounded near the stem, ⅜ inch broad. This is said to be rather rare. We found it twice in August growing solitary on the roadside in the grass. It was large-sized, measuring 7 inches across cap, of a grayish-white color, with prominent warts; the stem was mealy, the volva was large. It was marked with distinct rings near the base. When kept many hours the smell becomes disagreeable. The name is given on account of the shape of the warts, which are conspicuous.

AMANITA VAGINATA = a sheath.
The Sheathed Mushroom.

Cap gray, mouse color, sometimes slate-colored gray, and even brownish, 2 to 4 inches broad. It is thin and fragile, convex, and then nearly flat, with a slight mound or umbo, but sometimes none. It is deeply striate or grooved (sulcate) on the margin. **Stem** is white and often covered with mealy particles. It is slender, either hollow or stuffed, 3 to 5 inches long, ⅓ to ½ inch thick. It is not bulbous, but is sheathed quite high in a loose, soft wrapper, the remains of the volva. There is no ring. **Gills** are whitish, free from the stem, and rounded. It is easily broken. There are several varieties (Peck). In one the plant is white, Var. alba. In Var. livida the cap is a leaden brownish color, and in the Var. fulva the cap is tawny yellow and ochraceous. The mouse-colored form is the most common. We found many specimens in July and August.

CORTINARIUS CORRUGATUS = wrinkled.
The Wrinkled Cortinarius.

Cap gray, with a pinkish-yellowish tint, 2 inches broad, campanulate, sticky, broken up into squamules, pellicle scaling, margin thin. **Stem** slender, 5 inches long, shiny, mealy at apex, slightly bulbous. **Gills** gray color, adnexed, distant, ventricose. This is a pretty mushroom. The shade of color of the pileus is delicate. We found it in August in the woods.

BOLETUS FELLEUS = bitter.
The Bitter Boletus.

This Boletus varies much in color; our plant was a brownish-gray, a dingy color. **Cap** 3 to 8 inches broad, convex or nearly plane, glabrous, even, flesh white, turning to flesh or pink color when wounded. Taste bitter, tubes adnate, long, depressed around the stem, crowded. **Stem** variable, 2 to 4 inches long, about ½ to 1 inch thick, equal or tapering, reticulated above, bulbous or enlarged at base, a little paler than the pileus. The Boleti we found grew in great numbers, in different localities, and were of all sizes. The color of the reticulations was a brownish-gray.

BOLETUS GRISEUS = gray.
The Gray Boletus.

Cap dark gray, 2 to 4 inches broad, broadly convex, smooth, soft, silky, flesh whitish. Tubes adnate, slightly depressed, mouths small. **Stem** 2 to 4 inches long, 3 to 6 lines thick, yellowish, much reticulated, sometimes reddish toward the base. Our plant was of a brownish color at base, and grew in the month of September.

PSALLIOTA CAMPESTRIS = a field.
The Common Mushroom.

There are several edible species of the genus Psalliota, chiefly the Field or Common Mushroom, which is constantly seen on our tables. **Cap** varies from white and gray to brown. It is 2 to 4 inches broad, fleshy, convex, then flattened, dry, sometimes covered with silky fibrils, and when old smooth. The margin of the cap generally extends beyond the gills. Flesh white. **Stem** rather short, 1 to 3 inches long, ⅓ to ⅔ inch thick, white or whitish, slender, stuffed and then hollow, nearly even. **Ring** distant, simple. **Gills** free, ventricose, narrowing at both ends, thin, first a pink color, then afterward brown or blackish-brown. It grows in rich pastures or in meadows, and is found in autumn. It has a most delicious flavor.

AGARICUS PLACOMYCES.
The Flat-capped Mushroom.

Cap a whitish-gray, about 3 inches broad, convex, and then expanded and flat. It is covered with small, distinct, brown, persistent scales, except on the disc, where they are so close together that they appear of a brown color. **Stem** is long and slender, 3 inches and more, stuffed and then hollow, equal and bulbous at the base. It is whitish, but sometimes has yellowish stains toward the base. **Gills** are first white, then pink, and lastly a blackish-brown. It grows under trees, and is found in summer and autumn.

COPRINTUS ATRAMENTARIUS = ink.
The Inky Coprinus.

Cap gray or grayish-brown, smooth, except a slight scaly appearance on the disc. It is silky near the margin, and the margin is irregular. When young it is often egg-shaped. **Gills** crowded, whitish, soon becoming brown and then deliquescent. **Stem** smooth, hollow, white. It grows in clusters until late in the autumn. We found our plants on a lawn in great profusion in the month of October.

PLUTEUS CERVINUS = a deer.
The Fawn-colored Pluteus.

Cap about 3 inches broad, whitish-gray color, at first bell-shaped, then expanded, smooth, even, but afterward broken up into fibrils, margin entire; flesh soft, white. **Stem** 3 to 6 inches long, nearly equal and solid, whitish, striate with black fibrils. **Gills** rounded behind, free, crowded, ventricose, white, then flesh color as the spores mature. This is a common species, appearing early in the season—April to November. It usually grows from stumps and old logs. It can be easily known by its gills, being quite free from the stem, where it joins the pileus.

MUSHROOMS WITH A GREEN COLORED CAP

RUSSULA VIRESCENS = green.
The Greenish Russula.

Cap of a grayish-green color. It is 2 to 4 inches broad, dry and broken up into small warts, the margin straight, obtuse, even; flesh white. **Stem** 2 inches long and ½ inch thick, solid, spongy inside, firm, white, sometimes marked with lines (rivulose.) **Gills** free, whitish, narrowed toward the stem, somewhat crowded, sometimes equal and forked, with a few shorter ones between. It is easily distinguished by the dull green pileus, being without a cuticle, and scaly in the form of patches. It is found in woods in July and September. We have not seen a specimen of R. virescens, so have used Stevenson's description. Edible, taste mild.

RUSSULA FURCATA = a fork.
The Forked Russula.

Cap from 3 to 5 inches broad, of an olive green color, sometimes greenish umber, covered with a silky bloom, fleshy, gibbous, then plano-depressed and funnel-shaped, cuticle here and there separable; margin at first inflexed, then spreading. Flesh firm, thick, white. **Stem** 2 to 3 inches long, solid, firm, stout, white. **Gills** adnato-decurrent, thick, distant, broad, narrowed at both ends, often forked, white. Our specimen was 5 inches broad, and the margin slightly striate, and when the cuticle was removed it was purplish underneath. It was found in August, in woods. Poisonous, taste bitter.

MUSHROOMS WITH WHITE COLORED CAP

AMANITA VIROSA = poison.
The Poisonous Amanita.

Cap shining white, from 2½ to 4 inches broad, fleshy, at first conical and acute, afterward bell-shaped and expanded, viscous in wet weather, shining when dry, margin even, sometimes unequal, spreading and inflexed, flesh white. **Stem** 4 to 6 inches long, wholly stuffed, almost solid, split up into lengthwise fibrils, cylindrical from a bulbous base, surface torn into scales, springing from a loose, thick, wide volva which bursts open at apex. **Ring** large, loose, silky, splitting into pieces. **Gills** free, thin, a little broader toward margin, crowded, not decurrent, though the stem is sometimes striate. This is a poisonous species, but striking in appearance from the shining white of the whole fungus. Found in the woods in August.

AMANITA PHALLOIDES = appearance, phallus-like.
The Death Cup.

This species is considered the most deadly of all the poisonous mushrooms, and yet it is one of the most beautiful. We place it in the section of white-colored mushrooms, though the cap is sometimes tinged with light yellow and delicate green. **Cap** 2 to 4 inches broad, ovate, campanulate, then spreading, obtuse, with a cuticle, sticky in moist weather, rarely sprinkled with one or two fragments of the volva, the margin regular, even. **Stem** 3 to 5 inches long, ½ inch thick, solid, bulbous and tapering upward, smooth, white. **Ring**superior, reflexed, slightly striate, swollen, white. Volva more or less buried in the ground, bursting open in a torn manner at the apex, with a loose border. **Gills** free, ventricose, 4 lines broad, shining white. This species, as well as A. virosa, has a fetid odor when kept. We found it oftener than any other species of Amanita.

AMANITA NITIDA = to shine.
The Shining Amanita.

Cap whitish, 3 to 4 inches broad, somewhat compact, at first hemispherical, covered with angular, adhering warts, which become a dark color (fuscous.) It is dry, shining, the margin even; flesh white. **Stem** 3 inches long, 1 inch thick, solid, firm, with a bulb-shaped base, scaly, white. **Ring** superior, thin, torn, slightly striate, covered with soft weak hairs beneath, which at length disappear. **Gills** free, crowded, wide, nearly ½ inch broad, ventricose, shining white. This was also found in August. There is nothing more beautiful than these white poisonous Amanitas.

LEPIOTA NAUCINOIDES = a nut shell.
The Smooth Lepiota.

Cap a clear white, with sometimes a brownish tint on the disc, 2 to 4 inches broad, smooth. **Stem** 1 to 3 inches long, ¼ to ⅓ inch thick, growing thicker toward the base, as if it had a bulb, white, hollow, but stuffed with a cottony pith. **Gills** white, when old they assume a pinkish-brownish hue. **Ring** has a thick, external edge, but its inner edge is so thin that it often breaks from the stem and becomes movable. It is found in the fields, by roadsides, or in the woods, from August to November. We have not seen a specimen of this mushroom, which is said to be nearly equal to the common mushroom in edible qualities. It is considered to resemble it also in appearance, but Professor Peck says the different color of the gills when the plants are both young will distinguish them, and the thin collar and stuffed stem of

L. naucinoides is also different from thick-edged ring and hollow stem of A. campestris. (Psalliota.)

LACTARIUS PIPERATUS = peppery.
The Peppery Lactarius.

Cap white, 4 to 9 inches broad, fleshy, rigid, depressed in centre when young, reflexed margin, at first involute, when full grown the surface becomes funnel-shaped and regular, even, smooth, without zones; flesh white. **Stem** 1 to 2 inches long, 1 to 2 inches thick, solid, obese, equal or obconical, slightly covered with powder (pruinose), white. **Gills**decurrent, crowded, narrow, scarcely broader than one line, obtuse at edge, regularly dividing by pairs from below upward (dichotomous), curved like a bow (arcuate), then all extended upward in a straight line, white, with occasional yellow spots. The milk white, unchangeable, plentiful, and acrid. This is common in woods. The cap in one of our specimens turned yellow when old, and was slightly striate at the margin; it was dry and thick and had no odor. The flesh had a whitish-brownish tinge where the cuticle was peeled off. Found it *only* in August.

LACTARIUS VELLEREUS = fleece.
The Fleecy Lactarius.

Cap white, 5 to 7 inches broad, fleshy, compact, convex, saucer-shaped, the margin for a long time sloping downward, with short, downy hairs (pubescent), dry, zoneless. **Stem** 2 to 3 inches long, 1 to 1½ inch thick, stout, solid, equal, covered with innate, thin pubescence. **Gills** arcuate, adnato-decurrent, rather thick, acute at the edge, somewhat distant, rather broad, connected by branches, pallid, watery, white. Milk scanty, white, very bitter. It is not said to be edible. The cap tends to become a pallid, reddish tan. This description is partially taken from Stevenson. The specimen we found had the margin revolute, it was 2½ inches broad, and the stem 2 inches long. The flesh was white and the cap was turning a brownish color. The stem slightly tapered toward the base. The milk was scanty and peppery. Found in the beginning of August in the woods. It resembles L. piperatus.

BOLETUS ALBUS = white.
The White Boletus.

Cap white, from 1½ to 3 inches broad, convex, viscid when moist, flesh white or yellowish, tubes small, nearly round (subrotund), adnate, whitish, becoming ochraceous. **Stem** 1½ to 3 inches long, 3 to 5 lines thick, equal, white, sometimes tinged with pink near the base. We found only one specimen of the white Boletus in August. It grew in the woods. The flesh became yellow and the stem was 1¼ inch long, and it slightly tapered toward the base.

PLEUROTUS ULMARIUS = elm.
The Elm Pleurotus.

The word pleurotus is taken from two Greek words, meaning a side and an ear. It is given on account of the stem growing in a lateral or eccentric manner. The Elm Pleurotus, so called from growing on elm trees, is considered edible. Our specimen had the *cap* whitish, but stained in the centre with a rusty yellowish color, 3 to 5 inches broad, thick, firm, smooth, convex, then plane. The skin was cracked in a tessellated manner. Flesh was firm and white. **Stem** white, 2 to 4 inches long, 1½ to ¾ inch thick, firm, smooth, a little hairy at the base, and attached eccentrically to the cap. **Gills** white with a yellow hue, broad, rounded near the stem, slightly adnexed and not crowded. It was found in October, and is not common.

PLEUROTUS SAPIDUS = agreeable to taste.
The Palatable Pleurotus.

This species generally grows in clusters with the stem united at the base. Our specimen grew on a maple tree. The plants protruded from a large crack in the trunk of a tree, about four feet above the ground, and grew one above the other. They had not attained their full growth. During former seasons they had been seen of a large size. **Pileus** is from 2 to 5 inches broad, grayish-white, smooth. **Caps** often overlap one another. Flesh is white. Gills broad, whitish, decurrent, and often slightly connected by oblique branches. **Stem** is generally short and lateral. It grew in October. Professor Peck says that in edible qualities it resembles the oyster mushroom, P. ostreatus.

MUSHROOMS WITH THE CAP BROWN AND VARIOUS SHADES OF BROWN

CORTINARIUS CINNAMOMEUS = cinnamon.
The Cinnamon-colored Cortinarius.

Cap a golden brown or bright cinnamon color, 1½ to 4 inches broad, umbonate, silky, shining, squamulose, with yellowish fibrils, and then smooth. **Stem** 2 inches long, stuffed and then hollow, thin, equal, tapering toward the base, yellowish color, as also are the flesh and the veil. **Gills** adnate, broad, crowded, shining reddish-brown color. Our specimen had beautiful reddish-colored gills, Var. semisanguineus (Peck). It grows in woods from August to November.

COLLYBIA ACERVATA = a heap.
The Tufted Collybia.

The name of the species is derived from a Latin word meaning a heap, so called from the habit of growth. (Stevenson.) **Cap** tan brown color, 2 to 3 inches broad, flesh color when moist, whitish when dry, convex, then flattened, obtuse or gibbous, margin at first involute, then flattened and slightly striate. **Stem** 2 to 4 inches long, 1 to 2 lines thick, very hollow (fistulose), rigid, fragile, slightly tapering upward, rarely compressed, very smooth, except the base, even, color brown or reddish-brown. **Gills** are at first adnexed, soon free, crowded, linear, narrow, plane, flesh color and then whitish. It grows in tufts (cæspitose). The stems are sometimes white, tomentose at the base. Stevenson says the cap is flesh color, but our specimen was of a pale or tan brown color, less than 2 inches broad; when moist it was much paler. Found in mixed woods in September.

Psathyrella disseminata.
Photographed by C. G. Lloyd.

PSATHYRELLA DISSEMINATA = scattered.
The Widely-spread Psathyrella.

Cap a light-colored yellowish-brown, changing into an ash color; the disc with a yellowish shade; of an oval shape, then bell-shaped, and marked with lines, almost sulcate. The margin does not extend beyond the gills. It is a small mushroom, measuring from 2 or 3 lines across the cap to 1 inch. **Stem** about 1 inch long or more, fragile, hollow, sometimes curved and bending, smooth and light-colored. **Gills** adnate, rather broad, slightly narrowed at both ends, at first whitish and then turning a brownish color. The plants vary greatly in height and size, are sometimes cæspitose and at other times scattered. The disc in some specimens was slightly raised in the middle, almost umbonate. It was found about stumps and on the ground, at the end of May, in mixed woods. It soon withers, but does not melt into fluid.

HYPHOLOMA CAPNOIDES = smoke.
The Gray-gilled Mushroom.

Cap is reddish-brown, 1 to 3 inches broad, fleshy, convex, then flattened, obtuse, dry, smooth. The margin in our specimen was slightly revolute. Flesh white. **Stem** 2 to 3 inches long, 2 to 4 lines thick, growing together at the base (connate), hollow, equal, often curved, becoming silky, even, whitish at apex, and here and there striate. **Gills** gray color, adnate, easily separating, rather broad, waxy. The name is given on account of the smoke-colored gills. It is not common, and is generally found on or about stumps in the autumn.

HYPHOLOMA PERPLEXUM = perplexing.
The Perplexing Hypholoma.

Cap brownish and turning to yellow, 1 to 3 inches broad and slightly umbonate, flesh whitish. **Stem** nearly equal, 2 to 3 inches long, 2 to 4 lines thick, firm, hollow, slightly fibrillose, whitish or yellowish above, reddish-brown below. **Gills** thin, close, slightly rounded at inner end, at first pale yellow, then tinged with green, finally purplish-brown. Taste mild. It grows in clusters. We found it both on and around old stumps, in the woods. It is sometimes solitary. (Edible.)

COLLYBIA DRYOPHILA = oak-loving.
The Oak-loving Collybia.

Cap tan color, often varying in color, ½ inch broad, thin, convex, nearly plane, sometimes with margin elevated, irregular, smooth, flesh white. **Stem** equal or thickened at base, 1 to 2 inches long, 1 to 2 lines broad, cartilaginous, smooth, hollow, yellowish, or reddish like the cap. **Gills** narrow, crowded, adnexed or nearly free, whitish. This little mushroom we found in a thick woods late in September, growing among dead leaves. There were oak trees all around and a great many pines. The weather had been rainy, and it was pale-colored and looked water-soaked.

TRICHOLOMA IMBRICATA = a tile.
The Imbricated Tricholoma.

Cap reddish-brown, 3 inches broad, thick, fleshy, broadly convex, and then flattened, obtuse, dry, continuous at disc, but torn into scales and fibrillose toward the margin; flesh firm, white. **Stem** solid, stout, sometimes short, and conico-bulbous, 1½ to 2 inches long, and as much as 1 inch thick, sometimes longer and almost equal; white at apex. **Gills** slightly emarginate, almost adnate, somewhat crowded, about 3 inches broad, wholly white when young, at length reddish. It grows either scattered or in groups. It is found in pine woods in September and November.

BOLETUS ORNATIPES = ornate and foot.
The Ornate-stemmed Boletus.

Cap 2 to 5 inches broad, yellowish-brown, convex, dry, firm, glabrous or minutely tomentose, flesh yellow or pale yellow. **Tubes** adnate, plane or concave, the mouths small or middle size, a clear yellow. **Stem** 2 to 4 inches long, 4 to 6 lines broad, subequal, distinctly and beautifully reticulated, yellow without and within. In woods and open places.

BOLETUS BREVIPES = short and foot.
The Short-stemmed Boletus.

Cap dark chestnut color, 1½ to 2½ inches broad, thick, convex, covered with a tough gluten, margin inflexed, flesh white or yellowish. **Tubes** short, nearly plane, adnate, or slightly depressed around the stem, small, white and afterward dingy ochraceous. **Stem** ½ to 1 inch long, 3 to 5 lines thick, whitish, very short, not dotted, or rarely with a few inconspicuous dots at the edge. This plant was found in October, and looked as if it rested upon the ground, the stem was so short; the cap was covered with gluten.

Lepiota procera
Photographed by C. G. Lloyd

LEPIOTA PROCERA = tall.
The Tall Lepiota.

Cap reddish-brown, 3 to 6 inches broad, fleshy; when young egg-shaped, and then campanulate, and flattening out with a broad, 121obtuse umbo. The cuticle breaks up into brownish scales, close near the centre, but sometimes wanting at the margin. The centre or umbo is darker colored; flesh dry, tough and white. **Stem** ½ inch thick, and 5 to 10 inches long; it is straight or a little bent, swollen or bulbous at base, sometimes variegated with brownish scales; deeply sunk at apex into the cup of the pileus; hollow or stuffed. **Ring** distinct from the stem, continuous with cuticle of pileus when young. It becomes free when the cap is expanded, and is then movable and persistent. **Gills** far remote from the stem, with a broad plano-depressed cartilaginous collar, crowded, ventricose, broader in front, soft, whitish, sometimes becoming dusky at the edge. The gills vary in color. This mushroom is a handsome species and is quite common in woods and pastures. (Edible.)

BOLETUS EDULIS = edible.
The Edible Boletus.

Cap varies sometimes in color (our specimen was brown). It is often a tawny light brown, paler at the margin, 4 to 6 inches broad, flesh white or yellowish, tinged with red under the cuticle. **Tubes** convex, nearly free, long, **minute**, round, white, then yellow and greenish. **Stem** 2 to 6 inches long, 6 to 18 lines thick, straight or bending, subequal or bulbous, short, more or less reticulated, especially above, whitish, pale reddish or brown. Found in August. Our specimen was small, the stem only 1½ inch long. (Edible.)

BOLETUS SCABER = rough.
The Scabrous-stemmed Boletus.

Cap varies in color, 1 to 5 inches broad, yellowish tan color, smooth, viscid when moist, at length rivulose. Tubes free, convex, white, then dingy color, mouths of tubes very small and round. **Stem** 3 to 5 inches long, 3 to 8 lines thick, solid, tapering above, roughened with fibrous scales. We found two or three varieties of this Boletus, which seems to grow everywhere in great abundance, in summer and autumn, in woods and in open places. One variety was of a yellowish tan color, Var. alutaceus, in another the flesh changed slightly to pinkish when wounded, Var. mutabilis (Peck). (Edible.)

BOLETUS CASTANEUS = chestnut.
The Chestnut Boletus.

Cap a chestnut color, brown or reddish brown, 1½ to 3 inches broad, convex, nearly plane or depressed, firm, even, dry, minutely velvety (tomentose), flesh white. **Tubes** free, short, small, white, becoming yellow. **Stem** 1 to 2½ inches long, 3 to

5 lines thick, equal or tapering upward, even, stuffed or hollow, colored like the cap. This is one of the prettiest of the Boleti. The bright chestnut color of the pileus forms a contrast with the white tubes, and makes it striking in appearance. We found it on several occasions, as it is common in woods. There are differences of opinion in regard to its being edible.

BOLETUS CHRYSENTERON = golden.
The Golden Flesh Boletus.

Cap dark brown or reddish-brown, 1 to 3 inches broad, convex or plane, soft, covered with woolly scales, sometimes marked with lines, flesh yellow, red beneath the cuticle, often slowly changing to blue when wounded, mouths large, angular, unequal. **Stem** 1 to 3 inches long, 3 to 6 lines thick, rigid, fibrous, striate, equal, reddish or pale yellow. This species is variable. We found one where the flesh was white, another where the tubes changed finally to green, and one that had an olive tint in the pileus.

BOLETUS ILLUDENS = deceiving.
The Deceiving Boletus.

Cap yellow or olive brown, 3 inches broad, plane, dry, marked with areoles, that is, the surface is broken up into little areas or patches. Flesh thick, white, red under cuticle. **Tubes**greenish-yellow, turning dark green, adnato-decurrent, that is, broadly attached to the stem and running down it, ⅛ inch long. **Stem** 2½ inches long, stuffed with brownish fibres, reticulated near apex, paler color than cap, curved.

BOLETUS PACHYPUS = thick.
The Thick-stemmed Boletus.

Cap tan color, 4 to 8 inches broad, convex, somewhat covered with long, soft hairs pressed closely to surface, subtomentose; flesh thick, whitish, changing slightly to blue. **Tubes** rather long, depressed around the stem, mouths round, pale yellow, at length tinged with green. **Stem** 2 to 4 inches long, thick, firm, reticulated, at first ovate, bulbous, then lengthened, equal, tinted pale yellow and red. The stem in the specimen was ¼ inch thick, swelling from apex downward, but it often measures 2 inches in thickness. This Boletus is considered poisonous.

BOLETUS SUBTOMENTOSUS = almost velvety.
The Yellow-cracked Boletus.

Cap dark brown, 1 to 4 inches broad, convex or nearly plane, soft, dry, covered with soft, weak, appressed hairs, almost olivaceous, of the same color beneath the cuticle, often marked with cracks and divided into little patches; flesh white or pallid. **Tubes** adnate, or depressed around the **stem**, yellow, mouths large, angular. **Stem** 1 to 2½ inches long, 2 to 5 lines thick, stout, somewhat ribbed, or scurfy, with minute dots. The cap varies in color, it may be yellowish-brown. We found the dark brown species growing on decaying wood, in pine woods, during the month of September.

BOLETUS PIPERATUS = peppery.
The Peppery Boletus.

Cap reddish-brown or ochraceous, 1 to 3 inches broad, convex or nearly plane, smooth, slightly viscid when moist, flesh white or yellowish, taste acrid, peppery. **Tubes** long, large, unequal, plane or convex, adnate or nearly decurrent, reddish, ferruginous. **Stem** 1½ to 3 inches long, 2 to 4 lines thick, slender, almost equal, tawny yellow; at the base a bright yellow. The cap in our specimen was marked with cracks and patches, and the margin obtuse. The stem was rather curved, and the same color as the cap. Flesh yellow. Tubes a dark-reddish, decided color, which makes it a striking-looking mushroom. Taste peppery.

BOLETUS SORDIDUS = dingy.
The Dingy-colored Boletus.

Cap a dingy, dark brown, about 2 inches broad, flesh white, tinged with red. **Tubes** long, nearly free, ⅜ inch long, white, turning a dark bluish-green. **Stem** tapering toward apex, 2½ inches long, curved, solid, ½ inch thick, brownish, marked with darker streaks. The mouths of tubes were angular, and the stem striate in our specimen. Found in the woods in August.

BOLETUS SUBLUTEUS = almost, and yellow.
The Small Yellow Boletus.

Cap brownish yellow, 1½ to 3 inches broad, convex or nearly plane, viscid or glutinous when moist, often obscurely streaked (virgate). Flesh whitish or dull yellowish. **Tubes** plane or convex, adnate, small, nearly round, yellow, becoming ochraceous. **Stem** 1½ to 2½ inches long, 2 to 4 lines thick, equal, slender, pale or yellowish, dotted above and below the ring with reddish, brownish, moist, or sticky dots (glandules). **Ring** almost soft, glutinous, at first concealing the tubes, then collapsing and forming a narrow whitish or brownish band around the stem. Our Boletus had a brownish ring. The cap was covered with a sticky, skin-like layer, called the pellicle or cuticle, both terms having the same meaning.

BOLETUS AFFINIS = related.
The Related Boletus.

Cap reddish-brown, fading to yellow, 2 to 4 inches broad, convex above and almost plane, nearly smooth, flesh white. **Tubes** plane or convex, adnate or slightly compressed around the stem, at first white and stuffed, then yellowish, turning to rusty ochraceous when wounded. **Stem** 1½ to 3 inches long, 4 to 8 lines thick, nearly equal, even, smooth, paler than the cap. Our specimen had a few yellowish spots on the cap, and is called Var. maculosus. (Edible.)

PAXILLUS LEPTOPUS = thin and a foot.
The Thin-stemmed Paxillus.

This is the only specimen of the genus Paxillus that we have found. There is another species, P. involutus, which Professor Peck says is edible. Stevenson says that P. leptopus is a remarkable species, that it is distinguished from P. involutus by having the gills simple at the base, not united by interlacing or transverse veins (anastomosing). **Cap** was a light brownish-yellow; it varies from 1½ to 3 inches in breadth, eccentric or lateral, depressed in the middle, dry, covered with dense down, soon torn into scales, which are a dingy yellow. Flesh yellow. **Stem** short, scarcely 1 inch, tapering downward, yellow inside. **Gills** decurrent, tense and straight, crowded, narrow, yellowish, then darker in color. It was growing on the ground in September.

1. Boletus edulis.
2. Hypholoma perplexum.
3. Marasmius rotula.
4. Calostoma cinnebarinus.

MUSHROOMS WITH PURPLE OR VIOLET-COLORED CAP

CORTINARIUS ALBO-VIOLACEOUS = white and violet. The Violet-colored Cortinarius.

Cap whitish-violet, 2 to 3 inches broad, fleshy, convex, broadly umbonate or gibbous, dry, beautifully silky and becoming even; flesh juicy, a bluish-white color. **Stem** 2 to 4 inches long, solid, firm, bulbous, club-shaped, ½ to 1 inch thick. It is, both outside and inside, of a whitish violet color, often fibrillose above, with the cortina, and sometimes with the white veil, in the form of a zone at the middle. **Gills** adnate, 2 to 3 lines broad, somewhat distant, slightly serrulated, of a peculiar ashy violaceous color, at length slightly cinnamon from the spores. It has no odor and the taste is insipid. We found this in the woods in the month of October, growing on dead leaves; a pretty fungus from the violet tints.

Chapter 5

DESCRIPTIONS OF SOME FAMILIAR MUSHROOMS WITHOUT REGARD TO COLOR

Here follows a list of fungi that we constantly see, but which cannot be classified by the color of the cap.

POLYPOREI, PORE-BEARING FUNGI

FISTULINA HEPATICA = liver.
The Beefsteak Fungus.

This species grows on trees, oaks or chestnuts, in hot weather. **Cap** is of a dark-red color, which probably suggested the name. It is generally 2 to 6 inches broad, but often grows to an immense size. The surface is rough, the flesh thick, viscid above, soft when young, when old tough, covered with tenacious fibres. **Stem** short and thick. **Pores** at first pallid or yellowish-pink when young; they become brownish ochraceous when old. It is changeable in form, is sometimes sessile (without a stem), or it has a short lateral stem.

The genus Fistulina, to which this mushroom belongs, has the under surface of the cap covered with minute hollow pores, which are separate from one another and stand side by side.

The shape varies. It is sometimes long, shaped like a tongue, or roundish. It is peculiar-looking. It is considered good for food and nourishing, but the taste is said to be rather acid. The specimens we found varied from 2 to 5 inches in diameter. They were of a dark-red color, and were tough and old. They grew upon a tree in a large forest, and were not found anywhere else.

POLYPORUS BETULINUS = birch.
The Birch Polyporus.

We shall meet a great many fungi on our walks that belong to the genus Polyporus. They are generally leathery (coriaceous) fungi, and many grow on wood. A few are edible, but are not recommended as food. The species P. betulinus is found on living and dead birch trees. The specimens we found grew in great quantities, of all sizes, from 1½ to 6 inches broad. They were at first pure white, and then assumed a brownish tinge. The edges were obtuse, the caps fleshy, then corky, smooth, the upper ends not regular, oblique in the form of an umbo or little knob, the pellicles or outside layers thin and easily separated. Pores short, small, unequal, at length separating. The shape of the fungus is peculiar, a sort of semi-circular outline that may be called dimidiate. The margins were involute. They protruded from a split in the bark of a dead birch tree which lay prostrate on the ground, several feet in length, and it was literally covered with the fungi, some an inch wide and snow white, and the largest 5 or 6 inches in width, and of a brownish-gray tinge. These specimens became as hard as wood after they had been kept for some time. The thin skin peeled off easily and disclosed the snowy flesh beneath.

POLYPORUS PERENNIS = perennial.
The Perennial Polyporus.

Cap is cinnamon-colored, then of a date brown, leathery, tough, funnel-shaped, becoming smooth, zoned. **Pores** minute, angular, acute, at first sprinkled with a white bloom, then naked and torn. **Stem** slightly firm, thickened downward, velvety. This is a common species, and one meets with it everywhere on the ground, and on stumps, from July to January. The cap is 1½ to 2 inches broad, and the stem 1 inch long.

POLYPORUS PICIPES = pitch and foot.
The Black-stemmed Polyporus.

Cap pallid color, then turning chestnut, often a pale yellowish livid color, with the disc chestnut, fleshy, leathery, rigid, tough, even, smooth, depressed at disc or behind. Flesh white. **Stem** eccentric and lateral, equal, firm, at first velvety, then naked, and dotted black up to the pores. **Pores** decurrent, round, very small, rather slender, white, then slightly pale and yellowish. This fungus grows on the trunks of trees, and is found as late as the middle of winter.

POLYPORUS SULPHUREUS = brimstone.
The Sulphury Polyporus.

This mushroom gains its name from the color of its pores, which are of a bright sulphur color. It grows in tufted layers (cæspitose), sometimes 1 to 2 feet long, and it cannot be mistaken. **Cap** may measure 8 inches in breadth, and is of a reddish-yellow color, overlapping like the shingles of a roof (imbricated). It is wavy and rather smooth. Flesh light yellowish, then white, splitting open. **Pores** are minute, even, sulphur yellow. They retain their color much better than the pileus. The plants are generally without a stem, but there may be a short stem, which is lateral. They grow in clusters, all fastened together and one above the other, and of all sizes. We saw this fungus first in a dense woods, where its bright color at once attracted our notice. It was growing in a large cluster, closely packed one over the other. It is said to be good for food when young and tender.

POLYPORUS LUCIDUS = bright.
The Shining Polyporus.

One can never mistake this fungus. Its surface looks as if covered with varnish, rather wrinkled, a bright dark-red color, and its shape is varied and singular. We have seen it sometimes shaped like a fan, and like a lady's high comb, or in some fantastic form. Stevenson says it is a light yellow color and then becomes blood red chestnut. It is first corky, then woody. **Stem** lateral, equal, varnished, shining, of the same color as cap. **Pores** are long, very small, white and then cinnamon color. It grows on and about stumps during the summer. **Cap** is from 2 to 6 inches broad, and the stem 6 to 10 inches long, and 1 or more thick.

POLYPORUS VERSICOLOR = changeable.
The Changeable Polyporus.

This species is also common. It is found on dead wood, in all forms and colors. **Cap** variegated with different-colored zones; leathery, thin, rigid, depressed behind, becoming velvety. **Pores** minute, round, acute and torn, white, turning pale or yellow.

POLYPORUS ELEGANS = elegant.
The Elegant Polyporus.

Cap 2 to 4 inches broad, of one color, pallid, ochraceous or orange, shining, equally fleshy, and then hardened, becoming woody, flattened, even, smooth. Flesh white. **Stem** eccentric or lateral, even, smooth, pallid at first, abruptly black and rooting at the base. **Pores** plane, minute, somewhat round, yellowish-white, pallid. The cap differs in shape from others that have been described; it is not funnel-shaped nor streaked, and is scarcely depressed, and the flesh is thick to the margin. It grows on trunks of trees from July to November.

CLAVARIEI, OR CLUB-SHAPED FUNGI

We now come to another order, Clavariei, of which the first genus is Clavaria, from a word meaning a club. They are fleshy fungi, not coriaceous. They have no distinct stem and generally grow on the ground. We will mention a few of those we often see. They somewhat resemble coral in growth but not in color.

CLAVARIA STRICTA = to draw tight.
The Constricted Clavaria.

This Clavaria grows on trunks of trees. It is of a pale yellowish color, becoming a dusky brown (fuscous) when bruised. The base is about 3 lines long, thick and much branched. The branches and branchlets are tense and straight, crowded, adpressed and acute. Stevenson says that this species is uncommon in Great Britain.

CLAVARIA FLAVA = yellow.
The Pale Yellow Clavaria.

Stevenson does not mention this species, so it may be peculiar to this country. **Stem** is short and stout, thick, and abruptly dissolves into a dense mass of erect branches nearly parallel. The tips are yellow but fade when old. It branches below and the stems are whitish. Flesh white. It is recommended as well flavored and edible.

CLAVARIA PISTILLARIS = a pestle.
The Large Club Clavaria.

This species belongs to the largest of the unbranched kind. It is generally 3 to 5 inches high, and ½ to ⅔ of an inch thick at top. Light yellow color, then reddish, and dingy brown in decay. It is smooth and the flesh soft and white. It is rounded at the top and club-shaped. It tapers downward toward the base. Stevenson gives the height from 6 to 12 inches, but Professor Peck says he has not seen it as large in this country. It is found in open grassy places. It was late in the autumn when we discovered it. (Edible.)

CLAVARIA INEQUALIS = unequal.
The Unequal Clavaria.

This fungus is yellow and fragile. The clubs are alike in color, simple or forked, and variable. It is common in woods and pastures. We found it in September in the woods, rather wrinkled in appearance. It is not classed among the edible species.

TYPHULA = reed mace.

One may sometimes see among the dead leaves in the woods, minute slender bodies with thread-like stems, springing up from the ground, 2 to 3 inches high, of a white color and cylindrical in shape. They look like slender stems from which the blossoms have been plucked. They are called Typhula. They grow on dead leaves, on mosses, or on dead herbaceous stems. The name is taken from the Cat Tail family, the Typhaceae, which they somewhat resemble in miniature.

SCHIZOPHYLLUM COMMUNE = to split, a leaf and common.
The Common Schizophyllum.

There is but one species given by Stevenson of this genus, and, as the name demonstrates, it is common, at least in this country. In Great Britain it is rare. It grows on dead wood and logs. It has zones, either of gray or white color, and it is turned up at the edge (revolute). There is no flesh, and the pileus is dry. The gills are branched fan-wise. It is not a typical Agaric, but is more like some Polyporei. The gills are split longitudinally at the edge, and the two lips commonly turn backward (revolute).

HIRNEOLA AURICULA JUDAE.
The Jew's Ear.

There is one species belonging to the order Tremellodon that is quite common. It is called the Jew's ear. It is a very peculiar-looking fungus, shaped somewhat like the human ear, of all sizes, and grows in great quantities in the same place. It looks as if it were composed of a thick jelly, and becomes soft and tremulous when damp. Its color is dark, sometimes almost black. It is tough and cup-shaped, with ridges across it like an ear. The generic name, Hirneola, means a jug, and the specific name, Auricula Judae, a Jew's ear.

GASTEROMYCETES, OR STOMACH FUNGI

SCLERODERMA VULGARE = hard, skin, common.
The Common Hard-skinned Mushroom.

This species closely resembles the common potato in shape and color. It generally measures 2 to 3 inches across, and is of a pale brown color. It grows close on the earth, is folded toward the base, and firm in texture. The cuticle is covered with warts or scales.

CRUCIBULUM VULGARE = crucible, common.
The Common Crucible.

This little fungus is about ¼ of an inch across. It resembles a tiny bird's-nest with eggs in it. At first it looks like a cottony knot, closely covered; its apex is closed by a membrane, then its covering is thrown off, and the apparent tiny eggs are merely smaller envelopes, called the peridiola. These are lentil-shaped and pale, and are fastened to the inside of the covering by a long cord, which can be seen only through a strong lens.

CYATHUS VERNICOSUS = varnished.
The Varnished Cup.

This differs from the crucible in color, form and habitat. It is about ½ an inch high. It is bell-shaped, becoming broadly open like a trumpet, and of a slate or ash color. The mouth and lining shine as if varnished, and hence its name. The plants grow on the ground, on wood and on leaves.

LYCOPERDON CYATHIFORME = cup-shape.
The Cup-shaped Puff-ball.

This species of puff-ball is round with a contracted base. It is 4 to 10 inches across, a white or pinkish-brown color, afterward becoming a darker brown and covered with small patches. When the spores mature the upper part of the covering (peridium) becomes torn and only the lower part remains. It looks like a dark-colored cup with a ragged margin, and may be seen by the excursionist in the spring on the roadside. It has survived the winter frosts and storms. It is split and shabby looking. In August it is a whitish puff-ball, in the spring only a torn, brown cup.

LYCOPERDON PYRIFORME = pear-shape.
The Pear-shaped Puff-ball.

This species is shaped like a pear. It is from 1 to 4 inches high and is covered with persistent warts so small as to look like scales to the naked eye. It is of a dingy white or brownish-yellow. Its shape separates it from the puff-balls, especially from the warted puff-ball, L. gemmatum, which is nearly round with a base like a stem, an ashy-gray color, and the surface is also warty, but unequally so, and as the warts fall off they leave the puff-ball dotted. The pear-shaped puff-ball has little fibrous rootlets, and the plants grow in crowds on decaying trees.

GEASTER HYGROMETRICUS = moisture, measure.
The Wandering Earth Star.

This earth star is from 2 to 3½ inches wide. It is sessile, of a brownish color, and changes its form accordingly as the weather is moist or dry, hence the name. It is contracted and round in dry weather, and star-like in damp atmosphere, with its lobes stretched out on the earth. The covering consists of three layers, the two outermost split from the top into several acute divisions, which spread out like the points of a star. The innermost layer is round and attached by the base. There are one or more openings at the top for the escape of the spores.

PHALLUS IMPUDICUS = disgusting.
The Fetid Wood Witch.

In the first stages the plant is white, soft and heavy, in shape and size like a hen's egg. It is covered by three layers, the outer one firm, the middle one gelatinous, the third and inner one consists of a thin membrane. This phallus develops under the ground until its spores are mature. At length the apex is ruptured by the growth of the spore receptacle, and the stem expands and elongates, escaping through the top, and elevates the cap into the air. The stem at the early stage is composed of cells filled with a gluten. The stem afterward becomes open and spongy, owing to the drying of the gelatinous matter. The spores are immersed in a strong-smelling, olive-green gluten. They are on the outside of the cap and embedded in its ridges. A part of the volva remains as a sheath at the base of the stem. This plant develops so rapidly as to attain in a few hours the height of seven inches, the stem is of lace-like structure, pure white, and its appearance suggests the silicious sponge so ornamental in collections, commonly known as Venus' basket. The drooping cap is also lacey with a network, and the spores drip mucus and then dry up, in the meantime spreading around a carrion-like, fetid smell. The Phallus, therefore, differs greatly in appearance from the other genera of the order when it is seen above ground, but if one is successful in finding it at an early stage, under the surface of the earth, he will realize its relationship to the general group, and find it an interesting subject of study.

ASCOMYCETES, OR SPORE-SAC FUNGI

PEZIZA AUKANTIA = golden.
The Golden Peziza.

This species is 2 to 3 inches in diameter, its disc is bright orange color, while its exterior is pale and downy, owing to the presence of short, stout hairs. It is sessile or nearly so, and grows in tufts on the ground near stumps of trees. At first the disc is thin and brittle, with a raised margin, much waved, becoming incised, and finally spreads flat on the ground.

MORCHELLA ESCULENTA = food.
The Edible or Common Morel.

This is 2 to 4 inches high, stem about ½ inch in diameter. The cap is of a dull yellow color, olivaceous, darkening with age to a brownish tinge. It is oval-shaped, with dark hollows.

HELVELLA INFULA = name of a woollen head-dress.
The Cap-like Helvella.

This species is named Infula, because it is supposed to resemble in shape the sacred woollen head-dress worn by priests of Rome, by supplicants and victims, tied around the head by a ribbon or bandage, which hangs down on both sides. The stem is surmounted with a lobed cap, with two to four irregularly drooping lobes of reddish or cinnamon-brown color, and is about 3 inches in diameter. The stem is 2 or 3 inches high, usually smooth, but sometimes pitted. We found our specimen in the woods in August.

Cortinarius distans
Photographed by C. G. Lloyd

DIRECTIONS FOR USING KEYS

Let us suppose that the beginner finds a mushroom and wishes to name it. He has learned its component parts. He has remarked the names of the classes into which mushrooms are divided. How then shall he make use of the Keys? We will imagine that he has found a Cantharellus. The cap is yellow color, so let him turn to the list of fungi described under the section "Yellow and Orange," and see if it agrees in appearance with anyone of these. (It is necessary before consulting a key to find the color of the spores. This is done by cutting off the cap, and placing it, gills downward, on paper, and leaving it there for two or three hours. Having followed these directions in this case it will have been seen that the spores are white.)

After consulting the list of "Yellow and Orange" he will find that the first one mentioned is Cantharellus cibarius, the Chantarelle. The description resembles that of the mushroom found in every particular.

Now let the beginner go further, and prove the correctness of the name in another way. Turning to the section called "General Helps to the Memory," on page 68, and reading the names of the different genera under the headings until he comes to the name Cantharellus, he will find it in the table called "Mushrooms with gills running down the stems (decurrent)." This distinction is apparent in the specimen found. Again, let him turn to the list of white-spored Agarics, page 73, and he will find the name of the genus Cantharellus there. Now, as an additional test, let him turn to the key at the end of this work, the key to Hymenomycetes. He must have learned enough by this time to know that his mushroom belongs to this class, namely, the one that has spores produced upon the lower part of the cap, and, also, that it is an Agaric, from its having gills on the under side. Let him begin with Section A, "with cap." 1. Mushrooms with radiating gills beneath caps (Agarics). The key then follows: 1. Plants fleshy, soon decaying. 2. Turn to number 2. There are two descriptions, juice milky and juice watery; he will choose the second one, which is followed by the number 3. Then follows, stem central or nearly so; this agrees with the plant, and leads to 4. The first line reads "white spores," which is correct; then comes 5. There are four lines with descriptions, the last one, "no ring and no volva," is right, which leads to 7. There are here two lines belonging to 7, the second one, "gills in the form of folds, obtuse edge," is correct, and points to 10. This reads, "Gills decurrent, plant terrestrial, Cantharellus." The Key gives the name of the *genus* only. In the list of descriptions an attempt is made to mention some of the commonest species. These directions apply to all the keys alike.

Division I

Key to Hymenomycetes, Membrane Fungi

Hymenomycetes or membrane fungi are divided into two sections:

Section A, with cap.

Section B, without cap.

Section A is divided into four classes:

I. Mushrooms with radiating gills beneath caps, gill-bearing mushrooms (Agarics).

II. With pores or tubes beneath caps (Polyporei).

III. With spines or teeth beneath the cap or branches (Hydnei).

IV. Where the spore-bearing surface beneath the cap is even, smooth, or slightly wrinkled (Thelephorei).

Section B is divided into two classes:

I. Plants club-shaped and simple, or bush-like and branched (Clavariei).

II. Plants gelatinous and irregular (Tremellinei).

Section A

Class I Key to Gill-bearing Mushrooms (*Agarics*)

1.	Plants fleshy, soon decaying,	2.
	Plants leathery, woody, persistent,	12.
2.	Juice milky, white, or colored,	Lactarius.
	Juice watery,	3.
3.	Stem central, or nearly so,	4.
	Stem lateral, eccentric or wanting,	11.
4.	Spores white,	5.
	Spores rosy, pink or salmon color,	15.
	Spores yellowish-brown, ochre color,	17.
	Spores dark brown,	21.
	Spores black,	24.
5.	With volva and ring,	Amanita.
	Volva and no ring,	Amanita (sub-genus Amanitopsis).
	Ring and no volva,	6.
	No ring and no volva,	7.
6.	Gills free, ring movable, pileus scaly,	Lepiota.
	Gills adnate, pileus generally smooth,	Armillaria.
7.	Gills thin, edge acute,	8.
	Gills in the form of folds, obtuse edge,	10.

8.	Gills decurrent or stem fleshy.	Clitocybe.
	Gills sinuate, notched behind, stem fleshy,	Tricholoma.
	Gills adnate, not decurrent, stem cartilaginous,	Collybia.
	Stem fleshy, cap often bright color,	9.
9.	Plants rigid, gills even, cap bright,	Russula.
	Plants with waxy gills,	Hygrophorus.
10.	Gills decurrent, plant terrestrial,	Cantharellus.
11.	Spores white,	Pleurotus.
	Spores yellowish or brown,	Crepidotus.
12.	Gills serrated on their edges, stem central or lateral,	Lentinus.
	Gills entire, stem central,	13.
	Stem lateral or wanting,	14.
13.	Gills simple, pileus dry, soon withering, then reviving when moist,	Marasmius.
14.	Gills deeply splitting, with weak hairs,	Schizophyllum.
	Gills united by veins, plant corky,	Lenzites.
15.	Volva, no ring,	Volvaria.
	No volva, ring present,	Annularia.
	No volva, no ring,	16.
16.	Gills free, rounded behind, cohering at first,	Pluteus.
	Gills adnate or sinuate, stem fleshy, soft, waxy, cap fleshy, margin incurved,	Entoloma.
	Gills decurrent, stem fleshy,	Clitopilis.
17.	Ring continuous, pileus with scales,	Pholiota.
	Ring cobwebby or evanescent, not apparent in old specimens,	18.
	Ring wanting,	19.
	Stem with cartilaginous rind,	21.

18.	Gills adnate, plants on the ground,	Cortinarius.
19.	Gills decurrent, stem fleshy, gills easily separating,	Paxillus.
	Gills not decurrent, stem fleshy,	20.
20.	Pileus fibrillose, or silky,	Inocybe.
	Pileus smooth and sticky,	Hebeloma.
21.	Veil remaining attached to margin of pileus, often not seen in old specimens,	Hypholoma.
	Veil on stem as a ring,	22.
	Margin of cap incurved when young,	Naucoria.
22.	Gills separate on the stem,	Agaricus or Psalliota.
	Gills united with stem,	Stropharia.
	Gills adnate or sinuate,	23.
23.	Margin of pileus incurved when young,	Psilocybe.
	Margin of pileus always straight,	Psathyra.
24.	Pileus of normal form,	25.
25.	Pileus fleshy, membranaceous or deliquescent,	26.
26.	Gills deliquescent—inky fluid,	Coprinus.
	Gills not deliquescent —ring present,	Annellaria.
	Gills not decurrent—ring wanting,	27.
27.	Pileus striate—plants small,	Psathyrella.
	Pileus not striate, stem fleshy, margin exceeding the gills,	Panaeolus.

Class II Key to Pore-bearing Fungi (*Polyporei*)

1. Pores readily separating from cap, spores   Boletus.
    whitish or brownish,

2. Stems strictly lateral, pores in the form of tubes,   Fistulina.
    mouths are separate from each other (growing
    on wood),

3. Tubes not separable from each other, round,   Polyporus.
    angular, or torn, fleshy, leathery or woody,

(Key to species of Boleti may be found in Professor Peck's work on Boleti.)

Class III Key to Spine-bearing Fungi (*Hydnei*)

1. Spines awl-shaped, distinct at base,   Hydnum.

    Spines awl-shaped, equal; plant gelatinous,   Tremellodon.
    tremulous,

Class IV Key to Smooth Surface Fungi (*Thelephorei*)

1. Spores white, on ground, fleshy, tubiform,   Craterellus
    cap blackish, scaly, stem hollow,   Cornucopioides.

2. Coriaceous or woody, somewhat zoned,   Stereum.
    entire, definite in form,

Section B
Class I Key to Clavariei

1. Fleshy, branched or simple, without distinct   Clavaria.
    stem, growing on the ground,

2. Growing on trunks, yellowish, becoming dark,   C. stricta.
    much branched, tense and straight,

3. Yellow, stuffed, clubs simple or forked, of the   C. inequalis.
    same color,

4. Color changeable, becoming dark, light yellow,   C. pistillaris.
    then reddish, simple, fleshy, stuffed, obovate,
    clavate, obtuse,

Division II

Key to Gasteromycetes and Ascomycetes

Section A   Fungi that have the spores inside the cap. (Stomach fungi or Gasteromycetes.)

Section B   Fungi that have the spores in delicate sacs. (Spore sac fungi or Ascomycetes.)

Section A

1.	Fungi covered with a hard rind,	Scleroderma.
2.	In which the spores when ripe turn to dust,	4.
	Where spores are at first closed in a cup-like sac that resembles a bird's-nest,	3.
3.	Fungi with the outside covering bowl-shaped of one cottony layer,	Crucibulum, the Crucible.
	Outside covering tubular, trumpet-shaped, of 3 layers,	Cyathus, the cup.
	Outside covering opening with a torn mouth,	Nidularia, bird's-nest.
4.	Outer covering splitting into star-like points,	Geaster, earth star.
	Outer covering opening by a single mouth at the top,	Lycoperdon, puff-ball.
	Spores at first borne in an egg-like sac, when ripe elevated on a cap at the top of the stem, no veil, has an odious smell,	Phallus, stink-horn fungus.

Section B

1.	Where the sacs soon become free, no special covering, mostly fleshy, cup-like fungi,	Peziza, cup fungus.
	Sacs opening from the first, caps pitted or furrowed,	2.
2.	Cap lobed, irregular, saddle-shaped,	Helvella, yellowish fungus.
	Cap oval or conical, upper surface with deep pits formed by long ridges,	Morchella or Morel, honey-combed fungus.

(The genera described under Section B. all belong to the order of Discomycetes, fungi that have the spore sacs collected in a flattened disc.)

APPENDIX

A guide for the identification and differentiation of agarics, comprised in four tables, arranged with reference to the colors of the spores, viz.:

Table I.	White spores.
Table II.	Red and pink spores.
Table III.	Ochraceous spores.
Table IV.	Dark purple and black spores.

NOTE

In using this table the student should first ascertain the color of the spores of the specimen under investigation. This will determine the particular table to be applied to its further examination. If, for instance, he finds its spores to be white, he will know that Table I. is the one to be consulted. Turning to that table, he should recall the place of its growth, its habitat. Now, suppose it to have been found growing on a stump, he will, by looking at the first column, Habitat, of Table I., be informed that it must be one of the four genera named in the column with the heading "On Stumps." Let him then examine its "gills." If he finds them to be "adnate," he will be assured that it must be an "Armillaria," as no other genus is shown in the column as growing "on stumps" and which has gills that are adnate. But to make assurance doubly sure, he may proceed further to discover whether the specimen has also the ring called for in column headed "Ring." If it has, and was found growing in the summer, he may feel quite safe in classifying it as Armillaria. Sometimes the same genus will be found in more than one column. This ought not to mislead or confuse the beginner. In Table I., column headed "Volva," Amanita is mentioned, and also in the column headed "Ring," but this indicates that an Amanita has both the Volva (the universal veil) and the Ring. So in the columns headed by "Stem," Pleurotus is represented as having a lateral or eccentric stem, and also as having no stem. The meaning is, that some species of the genus have no stem, while there are others in which the stem is lateral or eccentric.

Transcriber's Note:
Variations in spelling, wording and table format are as in the original.

Table I   White Spores

Size of plants, small.		Collybia,[1] Mycena, Omphalia, Marasmius.
Plants deliquescent.		
Time of growth, summer.		Amanita, Collybia, Mycena, Omphalia, Lepiota, Pleurotus, Russula,[2] Lactarius.
Time of growth, autumn.		Amanita, Clitocybe, Collybia, Mycena, Omphalia, Hygrophorus, Lepiota, Marasmius, Armillaria, Pleurotus, Tricholoma, Russula, Cantharellus, Lactarius.[3]
Habitat	In woods, in uncultivated places, on ground.	Amanita, Armillaria, Tricholoma,[4] Clitocybe, Collybia,[5] Hygrophorus, Lactarius, Russula, Cantharellus.[6]

	In grass and fields, on ground.	Lepiota, Tricholoma.[7]
	On other plants— epiphytal.	Mycena, Omphalia, Marasmius, Collybia.
	On stumps.	Panus, Armillaria, Lenzites, Lentinus.
	On wood.	Trogia, Pleurotus, Schizophyllum,[8] Cantharellus.[9]
	On manure. [Category missing in original.]	
Gills,	free.	Amanita, Lepiota.
	adnate.	Armillaria, Clitocybe, Collybia.
	decurrent.	Omphalia, Clitocybe, Cantharellus, Hygrophorus, Lactarius.[10]
	serrated.	Lentinus.
	sinuous.	Tricholoma, Pleurotus.
	distant.	Marasmius, Clitocybe.
	in folds.	Cantharellus, Trogia.
Volva.		Amanita.
Veil adhering to margin of cap.		Tricholoma.
Ring.		Amanita, Armillaria, Lepiota.

Stem,	cartilaginous.	Marasmius, Mycena, Omphalia, Collybia.
	lateral, or eccentric.	Pleurotus, Panus.
	none.	Lenzites, Pleurotus, Trogia, Schizophyllum, Panus.
	brittle.	Russula.
Pileus,	scaly or warted.	Amanita, Lepiota.
	campanulate.	Mycena.
	silky, cracked or fibrillose.	Tricholoma, Clitocybe, Pleurotus.
	umbonate.	Mycena.
	umbilicate.	Omphalia, Lactarius.[11]
	striate.	Omphalia, Mycena.
Pileus and Gills milky.		Lactarius.

1. Some small.
2. In late summer.
3. Generally in autumn.
4. Large species.
5. Few.
6. Some.
7. Small species.
8. Sometimes on rotten wood.
9. Some on rotten wood.
10. Adnato decurrent.
11. Becomes depressed in centre.

Table II   Red and Pink Spores

Size of plants, small.		Leptonia.
Plants deliquescent.		
Time of growth, summer.		Volvaria, Pluteus, Enteloma, Leptonia, Nolanea, Eccilia.
Time of growth, autumn.		Volvaria, Pluteus, Nolanea,
Habitat	In woods, in uncultivated places, on ground.	Volvaria,[1] Enteloma, Clitopilus, Leptonia,[2] Nolanea,[3] Claudopus.
	In grass and fields, on ground.	Nolanea.
	On other plants—epiphytal.	
	On stumps.	Pluteus.[4]
	On wood.	Volvaria,[5] Claudopus.
	On manure.	
Gills,	free.	Nolanea, Pluteus, Annularia, Volvaria.
	adnate.	Nolanea, Enteloma.[6]
	decurrent.	Eccilia, Clitopilus, Claudopus.
	sinuous.	Enteloma, Claudopus.
	serrated.	
	distant.	
	in folds.	
Volva.		Volvaria.
Veil adhering to margin of cap.		Enteloma.
Ring.		Annularia.

Stem,	cartilaginous.	Nolanea, Leptonia.
	lateral, or eccentric.	Claudopus.
	none.	Claudopus.
	brittle.	
Pileus,	scaly or warted.	Leptonia.
	campanulate.	Leptonia, Nolanea.
	silky, cracked or fibrillose.	Entoloma, Pluteus.[7]
	umbonate.	Pluteus.[8]
	umbilicate.	Leptonia, Eccilia.
	striate.	Nolanea.
Pileus and Gills milky.		

1. Damp ground.
2. Dry hills.
3. Wet places in woods.
4. On or close to stumps.
5. On rotten wood.
6. Almost free.
7. Often fibrillose or floccose.
8. Somewhat.

Table III   Ochraceous Spores

Size of plants, small.		
Plants deliquescent.		
Time of growth, summer.		Pholiota, Inocybe, Naucoria.
Time of growth, autumn.		Inocybe, Flammula, Pholiota, Galera, Hebeloma, Crepedotus, Naucoria, Cortinarius.
Habitat	In woods, in uncultivated places, on ground.	Inocybe, Pholiota,[1] Hebeloma, Flammula, Paxillus, Cortinarius, Naucoria, Galera.

	In grass and fields, on ground.	Cortinarius.
	On other plants— epiphytal.	Naucoria.
	On stumps.	Pholiota, Paxillus.
	On wood.	Claudopus, Flammula, Crepidotus, Naucoria.
	On manure.	
Gills,	free.	Naucoria.
	adnate.	Naucoria, Pholiota,[2] Flammula, Cortinarius, Hebeloma.
	decurrent.	Flammula, Paxillus.
	sinuous.	Hebeloma.
	serrated.	
	distant.	
	in folds.	
Volva.		
Veil adhering to margin of cap.		Hebeloma, Cortinarius, Inocybe.
Ring.		Pholiota, Cortinarius.[3]
Stem,	cartilaginous.	Tubaria, Naucoria, Galera.
	lateral, or excentric.	Crepidotus.
	none.	Crepidotus.
	brittle.	

Pileus,	scaly or warted.	Flammula, Inocybe.
	campanulate.	Galera, Pluteolus.
	silky, cracked or fibrillose.	Inocybe.
	umbonate.	Inocybe.
	umbilicate.	
	striate.	Pluteolus, Galera.
Pileus and Gills milky.		

1. Damp ground.
2. Somewhat free.
3. Some with rings.

Table IV Dark Purple and Black Spores

Size of plants, small.		Psathyrella.
Plants deliquescent.		Coprinus, Bolbitius.
Time of growth, summer.		Coprinus, Stropharia, Panaeolus.
Time of growth, autumn.		Coprinus, Psaliota, Panaeolus, Hypholoma.
Habitat	In woods, in uncultivated places, on ground.	Stropharia, Psathyra.
	In grass and fields, on ground.	Psaliota.
	On other plants— epiphytal.	Stropharia.

	On stumps.	Hypholoma, Psathyra.
	On wood.	Psathyra,[1] Hypholoma.
	On manure.	Stropharia, Panaeolus, Psathyrella, Coprinus, Bolbitius.
Gills,	free.	Chetonia, Psalliota, Psathyrella, Coprinus, Bolbitius.
	adnate.	Stropharia, Hypholoma, Psathyrella.
	decurrent.	Gomphidius.
	sinuous.	Hypholoma.
	serrated.	
	distant.	
	in folds.	
Volva.		
Veil adhering to margin.		Hypholoma.
Ring.		Stropharia Psalliota, Gomphidius.[2]
Stem,	cartilaginous.	Psathyra, Psilocybe.
	lateral, or excentric.	
	none.	
	brittle.	

Pileus,	scaly or warted.	
	campanulate.	Psathyra, Psathyrella,[3] Coprinus, Gomphidius.[4]
	silky, cracked or fibrillose.	
	umbonate.	
	umbilicate.	
	striate.	Psathyra, Psathyrella.
Pileus and Gills milky.		

1. On rotten wood.
2. A floccose ring.
3. At first, adpressed to stem.
4. Top shaped.

Table IV Dark Purple and Black Spores (*Continued*)

Gills, in folds.	Volva.	Veil adhering.	Ring.
		Hypholoma	Stropharia, Psalliota, Gomphidius.[2]
Stem, Cartilaginous.	Stem, lateral, or eccentric.	Stem, none.	Stem, brittle.
Psathyra, Psilocybe			
Pileus, scaly or warted.	Pileus, campanulate Psathyra, Psathyrella,[3] Coprinus, Gomphidius.[4]	Pilous, silky, cracked or fibrillose.	Pileus, umbonate

Pileus, umbilicate.	Pileus, stristo.	Pileus and gills milky.	
	Psathyra, Psathyrella.		

2. A floccose ring.

3. At first adpressed to stem.

4. Top shaped.

GLOSSARY

Acute´ Gills when called acute have sharp edges or are pointed at either end.

Adnate´ Spoken of gills when they are firmly attached to the stem.

Adnex´ A less degree of attachment of gills than adnate.

A´garic A mushroom that bears gills.

Aluta´ceous A light leather color.

Anas´tomosing Interlacing of veins, spoken of gills that are united by cross veins or partitions.

An´nulus The ring on the stem of a mushroom, formed by the separation of the veil from the margin of the cap.

A´pex. The top The end of the stem nearest to the gills.

Ap´ical Relating to the apex.

Appendic´ulate Hanging in small fragments.

Arach´noid Like a cobweb.

Ar´cuate Shaped like a bow.

Are´olate Any surface divided into little areas or patches.

Axis Stipe or stalk.

Band A broad bar of color.

Basid´ium (plural basidia) Mother cells in the hymenium.

Behind Posterior, the end of a gill next to the stem is said to be the posterior end.

Bifur´cate Two-forked.

Bulbous Spoken of the stem when it has a bulb-like swelling at the base.

Caes´pitose Growing in tufts.

Campan´ulate Bell-shaped.

Cap The pileus.

Cartilag´inous Gristly, tough.

Casta´neus Chestnut color.

Cell A mass of protoplasm, with or without an enclosing wall.

Chlorophyll The green coloring-matter contained in plants.

Cla´vate Club-shaped.

Close Crowded together—term used in describing gills.

Cohe´rent Sticking together.

Con´cave Having a rounded inwardly curved surface.

Concen´tric With a common centre, as a series of rings, one within the other.

Con´nate Growing together from the first.

Constric´ted Contracted.

Contin´uous Without interruption.

Convex Elevated and regularly rounded.

Con´volute Covered with irregularities on the surface, like the human brain.

Coria´ceous Leathery in texture.

Cor´rugated Wrinkled.

Corti´na A veil of cobwebby texture. It gives the name to the genus Cortinarius.

Cre´nate In wavy scallops.

Cu´ticle Pellicle, a skin-like layer on the outside surface of the cap and stem.

Cy´athiform Cup-shaped.

Decid´uous Falling off when mature at the end of the season.

Decur´rent Gills that run down the stem are called decurrent.

Dehis´cence The opening of a peridium, when ripe, to discharge the spores.

Deliques´cent Turning to liquid when mature.

Dichot´omous Two-forked, regularly dividing by pairs from below upward.

Dimid´iate Divided into two equal parts, applied to gills that only reach half-way to the stem, and to the cap when it is semi-circular or nearly so.

Disc The central part of the upper surface of the cap.

Distant Gills when they are far apart.

Emar´ginate A gill which has a sudden curve in its margin close to the stem.

Entire An edge that is straight, has no notch.

Ep´iphytal Growing on the outside of another plant.

Equal A stem is equal when it is of uniform thickness, gills when they are of equal length.

Eccen´tric A stem which is not in the centre, but is attached to the cap between the margin and centre.

Fascic´ulate Growing in clusters.

Ferru´ginous Color of iron rust.

Fi´brous Composed of fibres.

Fis´tulose Tubular, hollow.

Fleshy Composed of juicy cellular tissue.

Floccose Woolly, downy.

Free Gills when not attached to the stem.

Fungus (plural Fungi) A plant that has no chlorophyll, and obtains its nourishment from dead or living organic matter.

Fus´cous Dingy dark-brown, or gray color,

Gelat´inous Of the nature of jelly.

Genus A number of species that have the same principal characteristics.

Gib´bous Swollen unequally— applied to the cap.

Gill Lamella, a radiating plate under the cap of an Agaric.

Gla´brous Smooth.

Glo´bose Nearly round.

Gran´ular Consisting of or covered with grains.

Grega´rions Growing in groups.

Hab´itat Place of growth.

Homoge´neous Of like nature.

Hyme´nium The fruit-bearing surface, a continuous layer of spore mother cells.

Hy´phae (singular Hypha) Elementary threads of a fungus, cylindrical, thread-like bodies, developing by growth at the apex.

Im´bricated Overlapping like the tiles of a roof.

Incras´sated Thickened.

Inferior Applied to a ring that is far down on the stem.

Infundibuliform Funnel-shaped.

Involute Rolled inward.

Labyrin´thine Like a labyrinth.

Lac´erate Torn.

Lamel´la See gill.

Line 1/12 of an inch.

Mac´ulate Spotted.

Me´dial or median When the ring is situated in the middle of the stem.

Membrana´ceous Thin, soft, like a membrane.

Mica´ceous Covered with shining particles, like mica.

Mother cell A cell from which another is derived.

Myce´lium The vegetative part of fungi, commonly called the spawn.

Mycol´ogist One who is versed in the study of fungi.

Obo´vate Having the broad end turned toward the top.

Ob´solete Nearly imperceptible.

Obtuse Blunt.

Ochra´ceous Light brownish-yellow.

Ovate Egg-shaped.

Par´asite A plant growing on another living body, from which it gains its nourishment.

Pel´licle See cuticle.

Peren´nial Growing from year to year.

Perid´ium The outer covering of the spores in some fungi, as in puff-balls.

Peridi´olum The inside peridium containing the spores.

Pi´leus See cap.

Pir´iform or pyriform Pear-shaped.

Plane Level surface.

Pores The tubes in Polyporei.

Poste´rior Term applied to the end of the gill next to the stem.

Pru´inose Covered with a bloom or powder.

Pulver´ulent Covered with powder or dust.

Putres´cent Decaying.

Rad´icating Taking root.

Retic´ulated Marked with cross lines like a net.

Rev´olute Rolled upward or backward.

Ri´mose Cracked.

Rim´ulose Covered with small cracks.

Ring Annulus.

Riv´ulose Marked with lines like rivers in maps.

Rotund´ Round.

Ru´gose Wrinkled.

Sap´id Agreeable to the taste.

Sap´rophyte A plant that lives on decaying matter.

Scab´rous Rough.

Scis´sile Easily split.

Sep´arating Spoken of gills when they easily separate from the stem.

Ses´sile Stemless.

Sin´uate Wavy, A gill that has a sudden curve near the stem.

Sor´did Dingy.

Spore The same body that answers to the seed of flowering plants.

Spo´rophore That part which bears the spores or spore mother cells.

Squa´mose Scaly.

Stalk A stipe or stem.

Stel´late Star-shaped.

Stipe See stalk.

Strobil´iform Shaped like a pine-cone.

Stuffed When a stem is filled with pith or a spongy substance.

Suc′culent   Juicy, fleshy.

Sul′cate   Grooved.

Supe′rior   Spoken of a ring that is high up on the stem.

Tes′sellated   In small squares, or checkered.

To′mentose   Covered with matted wool.

Tra′ma   The substance proceeding from and of like nature with the part that bears the hymenium—the framework of the gills.

Trem′elloid   Jelly-like.

Tu′bæform   Trumpet-shaped.

Umbil′icate   Having a central depression.

Um′bo   Arising or mound in the centre of the cap.

Veins   Swollen wrinkles on the sides and at the base between the gills.

Ven′tricose   Swelling in the middle.

Ver′nicose   Varnished.

Vil′lose   Covered with weak, soft hairs.

Vires′cent   Greenish.

Vir′gate   Streaked.

Vis′cid   Sticky.

Vis′cous   Gluey.

Zones   Circular bands of color.

INDEX